PERGAMON INTERNATIONAL LIBRARY
of Science, Technology, Engineering and Social Studies
The 1000-volume original paperback library in aid of education,
industrial training and the enjoyment of leisure
Publisher: Robert Maxwell, M.C.

WHY AND HOW
Some Problems and Methods
in Historical Biology

FOUNDATIONS & PHILOSOPHY OF SCIENCE & TECHNOLOGY
General Editor: MARIO BUNGE, *McGill University, Montreal, Canada*

This new series has three goals. The editor formulates them as follows:

1) To encourage the systematic exploration of the foundations of science and technology

2) To foster research into the epistemological, semantical, ontological and ethical dimensions of scientific and technological research

3) To keep scientists, technologists, scientific and technological administrators and policy makers informed about progress in the foundations and philosophy of science and technology

Some Titles in the Series

AGASSI, J.
The Philosophy of Technology

ANGEL, R.
Relativity: The Theory and its Philosophy

BUCHTEL, H.
The Conceptual Nervous System

BUNGE, M.
The Mind-Body Problem

HATCHER, W.
The Logical Foundations of Mathematics

WILDER, R.
Mathematics as a Cultural System

Other Titles of Interest

ADVANCES IN THE BIOSCIENCES*

This review journal offers a comprehensive collection of leading scientists' latest experimental work. It makes available the abstracts and discussion contributions made at symposia and workshop conferences. The newest developments and directions in biological research are outlined in the symposia; the workshop discussions offer the scientist engaged in experimental study valuable insight into the subjects, methods and results of the work being done by leading research teams.

* FREE SPECIMEN COPIES AVAILABLE UPON REQUEST

WHY AND HOW

Some Problems and Methods in Historical Biology

by

GEORGE GAYLORD SIMPSON

PERGAMON PRESS

OXFORD • NEW YORK • TORONTO • SYDNEY • PARIS • FRANKFURT

U.K.	Pergamon Press Ltd., Headington Hill Hall, Oxford OX3 0BW, England
U.S.A.	Pergamon Press Inc., Maxwell House, Fairview Park, Elmsford, New York 10523, U.S.A.
CANADA	Pergamon of Canada, Suite 104, 150 Consumers Road, Willowdale, Ontario M2J 1P9, Canada
AUSTRALIA	Pergamon Press (Aust.) Pty. Ltd., P.O. Box 544, Potts Point, N.S.W. 2011, Australia
FRANCE	Pergamon Press SARL, 24 rue des Ecoles, 75240 Paris, Cedex 05, France
FEDERAL REPUBLIC OF GERMANY	Pergamon Press GmbH, 6242 Kronberg-Taunus, Pferdstrasse 1, Federal Republic of Germany

First edition 1980

British Library Cataloguing in Publication Data

Simpson, George Gaylord
Why and how. — (Foundations and philosophy of
science & technology).
1. Vertebrates, Fossil
2. Paleontology — Methodology
I. Title II. Series
566'.01'8   QE841   79-42774

ISBN 0-08-025785-2 (Hardcover)
ISBN 0-08-025784-4 (Flexicover)

Printed in the United States of America

CONTENTS

Introduction

A FICTIVE character who did not consider himself intelligent is credited by his creator with the remark that "it is worth a lot of bother to be able to think properly." I consider that an intelligent remark, and it has frequently recurred to me as I partly assembled and partly wrote the present book. The book is based on gleanings from my past efforts to think properly over a great many years, here assembled, modified in some respects, and supplemented in others by continuing efforts to think properly.

The original idea, suggested by Professor Mario Bunge as editor of a series of books for Pergamon Press, was to assemble and reprint a number of my papers bearing on methods in biology not previously collected or published in book format. That has been done, but with some modification as the work proceeded. As to the field covered, it is better designated as historical biology. That is not a usual specification of a subdivision of the vast subject of biology, so it calls for some explanation.

I am generally designated as a paleontologist, more specifically as a vertebrate paleontologist, and most specifically as a paleomammalogist. Paleontology is definable as the study of fossils, and fossils are so associated with geology that paleontologists are often considered as one kind, perhaps as a rather odd kind, of geologist. However, paleontologists are also biologists, and some of them, like me, become quite as much concerned with biology as with geology. The particular aspects of biology that most occupy us are organismal and evolutionary: organismal because they deal with organisms as such and not, or not primarily, as congeries of cells or molecules; evolutionary because they deal with the course, causes, and principles of the evolution of whole organisms. The study of fossils is obviously relevant to those subjects, and when paleontology is combined with them the result is an overall approach appropriately designated as historical biology. It exemplifies that approach that this book begins with some aspects of the study of the fossil record and closes with a historical analysis of the now living fauna of a continent.

The papers here reprinted or excerpted all involve methods directly or indirectly, sometimes in an explicit and mechanical sense, more often as concepts and methods of thought and problem-solving. In most cases the methods are brought out and simultaneously exemplified by their application to some particular research, which is the "why" of the book's title as the "how" is the methodological aspect.

Many of the items here collected are excerpts from rather than full reprints of the original papers involved. Most of them have also been slightly modified to fit them into the present context. However, I have taken care not to introduce later information or ideas into the texts, which would have falsified the apparent state of knowledge at the time of original publication. In each case the text here given is followed by a bibliographic reference to its source. In two instances, as stated in comments on them, the present text is based on an earlier draft that differed somewhat from the previously published form.

Each of the six sections of this book begins with a general statement about the subjects to be covered. Each separate text is preceded by a comment, sometimes of considerable length, the purpose of which is to provide background and context for the following text and also, when appropriate, to discuss the present status of the subject. Thus although the book is centered on papers previously published, its comments are new and in sum present a current point of view on the subjects treated.

Although the subjects are quite diverse, they do not begin to cover all aspects of historical biology, even as regards the applicable methods. This work necessarily concentrates on subjects that I have earlier discussed in print and deem to have current interest. The fossil record, morphology, paleontology, taxonomy, biometry, and biogeography are subjects which I have constantly studied and on which I have often published, but, with one exception, to which I have not devoted a book. The exception is taxonomy, on which I did publish a book in 1961, but the material here presented in Chapter 4, on systematics and taxonomy, supplements or to some extent even supplants and does not repeat what was said in that book.

In most cases I have not repeated the full bibliographic documentation required in the original publications but not really necessary for the present purposes. Such citations as are necessary for those purposes are of course given. In several instances I have added references published after first publication of my paper and useful now for updating the subject.

The tables and figures have been renumbered for clearer reference in this collection. Many of the figure legends have been rewritten, usually to make them more nearly self-sufficient in the present less full context.

1

The Fossil Record

THE HISTORY of living things is an essential part of evolutionary biology. The most direct evidence bearing on this history consists of fossils, which may be defined as any remains or traces of organisms antedating human observation and recording. The basic, or one might say alpha level, data include the description of known fossils, but also considerably more. The localities where each was found, their associations with other fossils, and the composition, structure, and stratigraphy of the rocks containing them are also basic observational data. A second or beta level involves controlled inferences about the direct observations, and these lead in several different directions, of which two are most immediately necessary for any historical interpretation. One is the need to place any individual fossil or whole fossil community in a time frame, which may be sequential or may be approximated in terms of years. Another is the classification and naming of the fossils, a subject that involves principles and methods not confined to historical or even to evolutionary biology.

Much the greatest part of the literature on the fossil record is on those two levels. That need not be specially exemplified here, although all the subjects that are exemplified are based on a lifetime of work largely on those levels. They are the reasonably solid basis for any more interpretive, more theoretical, and ultimately more philosophical consideration of evolutionary biology. This chapter consists of a few examples of studies of the record as to its adequacy and as to ways of looking at and representing some of its aspects.

Bias, Adequacy, and Sampling of the Fossil Record

Comment

This is an extract from a longer paper read at an elaborate celebration at the University of Chicago in 1959 of the centenary of the publication of Darwin's *Origin of Species*. This extract is here reprinted in slightly modified form. There continues to be discussion of its subjects and some students of them have understood the adequacy and biases of the record other than in my way in 1959 or now. The next selection in this chapter is a briefer and more recent note bearing on that topic.

Subsequent discoveries of Pleistocene mammals in Florida have changed the figures given in Table 1–5, but this is still a valid example of a method.

Text

The main (still not *all*) causes of bias in the published fossil record as we have it may be outlined as follows:

(1) Biases of preservation
(A) From characteristics of the organisms
    (1) Differences in physical fossilizability, especially presence or absence of hard parts
    (2) Impossibility of complete preservation of any individual, and difference in parts preserved in different groups
    (3) Differences in habits and habitus affecting likelihood of fossilization
(B) From conditions of burial and subsequent preservation
    (1) Differences between the living community and its associated remains after death, transportation, and burial
    (2) Correlation of predominant sedimentation or erosion with different habitats and environments
    (3) Differential destruction after burial (e.g. by processes of preservation, metamorphism of the enclosing rock, or subsequent erosion)
    (4) Differential effects of intervals in sedimentation
(2) Biases of collection
(A) From requirement of recent exposure or near-exposure
    (1) Different chances of exposure of sediments of different environments
    (2) Limited and varying geographic distribution of exposed sediments of a given age or origin
    (3) Correlation of available facies and of extent of exposure with age
    (4) Correlation of effective exposure with present regional climates
(B) From the processes of discovery
    (1) Differences in accessibility of actual or potential fossil fields
    (2) Differences in total collecting effort
    (3) Differences in collecting methods
    (4) Differences in recognizability of fossils in the field
(C) From identification and other study
    (1) Differences in identifiability of preserved parts
    (2) Differences in recency and intensity of monographing
    (3) Individual idiosyncrasies of students, especially splitting or lumping

The factors thus listed under the name of "bias" are, for the most part, those usually considered (as those known to him were by Darwin) as causes for the "imperfection" or "incompleteness" of the record. This less usual approach has been made in order to stress the modern attitude and methodology of inference from samples and also the fact that bias and incompleteness are not the same. Incompleteness, as usually understood, has to do with the fraction of species (or other taxa) known among all that actually lived. Bias has to do with the fidelity with which a known sample represents a population in some defined respect. A sample with one specimen each of all the species of an ancient fauna or flora would be taxonomically complete, but it would be hopelessly biased with respect, for instance, to the relative abundance of those species. Similarly, it is evident that a highly incomplete sample could be unbiased for purposes of some particular inference.

Few concrete estimates of completeness have been made, and those few are subject to large uncertainties. They are, nevertheless, of interest and real value even if they indicate only a broad order of magnitude. Most general, but also least certain, is estimation of overall completeness for all organisms through all of the history of life. I (Simpson, 1952) have estimated the total number of species through all time as probably between 50 and 4,000 million, with 500 million as a reasonable single figure for magnitude. Cailleux (1954) estimated 17–860 million with a medial figure of 150 million, but his assumption of a constant geometric rate of increase seems to me almost certainly wrong and likely to give figures that are too low. By my figures (or guesses) perhaps 300 million species were animals. Muller and Campbell (1954) have estimated that about 1,000,000 living and few over 91,000 species of fossil animals have been described. If so, the proportion of extinct species now actually known may be about 0.03 per cent. In any event, it is probably on the order of hundredths and possibly even thousandths of 1 per cent: a discouraging conclusion!

This extreme incompleteness is, however, unduly influenced by a few classes (especially Insects) and phyla (especially the "worms" for which the record is exceptionally poor. Teichert (1956) has attempted an estimate of how many species are actually preserved as fossils, omitting the Precambrian because, whatever the reason, it does have few fossils and omitting the insects and soft-bodied invertebrates because of their comparatively poor chances of preservation. His total estimate is about 10 million, of which 7 or 8 million are presumed to be animals (including animal-like protists). He considers 10 million "at least 20, perhaps 50 times as many as the number of fossil species known to science." For animals Teichert's figures are more than 70 times the number of described species as carefully estimated by Muller and Campbell. Teichert apparently assumes that all species of animals with skeletons, except insects, have been preserved from the Cambrian onward and are available to collectors. That evidently exaggerated assumption must tend to make his figure decidedly too high. On the other hand, the soft-bodied invertebrates and insects, omitted by Teichert, are only comparatively rare as fossils. Their absolute numbers in the fossil record are large and may be expected to increase with more intensive collecting and the use of new methods. This tends to make Teichert's figure too low. Perhaps the two opposite errors tend to cancel out.

It is a guess, but a reasonably educated one, that on the order of 1 to 10 per cent of all species that ever lived are recoverable as fossils and that, of those, on the order of 1 to 10 per cent have so far been found and described. The task of describing recoverable species is no more than well begun, and when (if ever) it is finished, the record will still be very incomplete.

These estimates of the great incompleteness of the currently known record may give much too gloomy an impression of the usefulness of that record and of its adequacy for solving problems of evolution. The incompleteness must, to be sure, be kept in mind, but the present record is certainly highly useful and also quite adequate for many purposes. Knowledge of all species would not necessarily be much more useful than knowledge of a small sample of them. Indeed, for purposes of generalization and theory it would be impossible to deal with all species, and sampling would be essential. It does no harm that nature did the sampling, if we can estimate the kinds and extents of nature's biases.

In the first place, although the species is a basic unit and some evolutionary problems can be studied only at that level, there are many still more important problems of phylogeny, structural change, faunal development, and the like that can be studied just as well or even better in terms of genera or higher taxa. A small sample of species represents a much higher percentage of genera, still higher of families, and so on. Table 1–1 exemplifies this relationship. Continued intensive sampling from 1900 to 1941 more than doubled the number of species but increased the number of known genera by little more than half and that of known families by only one-eighth. The numbers of taxa originally present in these faunas is not known, but it is clear that a sampling very incomplete for species was decidedly more complete for genera and still more for families. This is not an extreme example; the discrepancy in recovery of species and higher taxa is frequently greater.

TABLE 1–1

Taxa of Fossil Mammals Known from White River Group
(Oligocene) in 1899 and 1941*
(Data Mainly from Scott and Jepsen, 1936–41, Modified)

	Families	Genera	Species
Known in 1941	39	95	228
Known in 1899			
Number	31	60	120
Per cent of 1941 total	79	63	47

* Taxa considered invalid in 1941, or in a few instances later, are not counted. Families and genera of which a species was known in 1899 are counted as known even if named later. Taxa known in 1899 but not from the White River group are counted as then unknown.

Another approach is by sampling experiments on faunas in which all or nearly all the taxa are known, hence necessarily recent faunas. Table 1–2 exemplifies such experiments for a relatively very small sample. It was postulated that probability of discovery was the same for all species, a false postulate but a legitimate simplification that is unlikely to affect the pertinent results seriously. Only 10 per cent of the species were selected at random, by numbering all the species and using a table of random numbers. That small sample, surely less complete than the samples of most well-worked fossil faunas, still included about a fifth of the genera, more than half the families, and almost two-thirds of the orders.

TABLE 1–2

Sampling Experiment on Land Mammals of New Guinea, Celebes,
and adjacent islands
(Based on List and Classification of Laurie and Hill, 1954)

	Orders	Families	Genera	Species
Total taxa in fauna	8	20	121	352
Taxa included in random				
sampling of 35 species:				
Number	5	11	23	35
Per cent	63	55	19	10

There is a related but distinct question as to the extent to which a local fauna may represent that of a larger region or a distant area. This is especially pertinent to paleontological sampling, in which the area available for collecting from beds of any one age is always far less than the area actually occupied by faunas of a given sort at that time. In this respect, too, there is evidence that local samples may be much more adequate than might appear at first sight, especially above the level of species. Table 1–3 is a somewhat extreme example from fossil faunas of the same age. The faunas compared are in the same broad category (land mammals) but are quite different in representation of included facies, as indicated by the small number of species (only seven) in common and by more detailed ecological analysis. Yet knowledge of the larger fauna, alone, would include 70 per cent of the total genera, 89 per cent of the families, and all the orders. Both samples are surely incomplete, but renewed planned intensive sampling should increase the percentage of representation from either fauna alone: it happens that the facies best represented in collections from Montana is at present least represented in those from New Mexico and vice versa.

TABLE 1–3

Comparison of Middle Paleocene Mammals of New Mexico and Montana
(Data Mainly from Matthew, 1937, and Simpson, 1937,
with Later Modifications)

	Orders	Families	Genera	Species
New Mexico				
Number	7	15	32	61
Per cent	100	79	56	52
Montana				
Number	7	17	40	63
Per cent	100	89	70	54
Total	7	19	57	117

Another approach to this aspect of sampling can, again, be based on completely known recent faunas. An example is given in Table 1–4. This example is also rather extreme, for the two areas compared, although in the same broad faunal region, are extremely different in climate, topography, and general ecological conditions. Yet the larger fauna includes the great majority of the total genera and almost all the families, and even the much smaller fauna has 43 per cent of the genera and 65 percent of the families.

TABLE 1–4

*Comparison of Recent Land (Non—volant) Mammals of
Florida and New Mexico*
(Data from Simpson, 1936)

	Orders	Families	Genera	Species
Florida				
Number	5	13	27	33
Per cent	83	65	43	21
New Mexico				
Number	6	19	54	133
Per cent	100	95	86	83
Total	6	20	63	160

Given the fact that the overall sampling of past life is and must always be highly incomplete, the fact that samples of various groups, times, and places vary so tremendously is actually an advantage, making the record more adequate for our purposes. At the two extremes, some desirable samples simply do not exist, but others are nearly complete. A rough estimate, at least, can often be made of the adequacy of a particular sample, and historical problems can be studied by examples based on the adequate available evidence. The probable reliability of conclusions can also be judged within increasingly close limits.

One approach to judgment of the taxonomic adequacy of samples is the rate of discovery and the intensity of exploration. This matter has been well discussed by Newell (1959) and needs only brief mention here. Recent mammals and birds, for instance, have been collected practically everywhere that they occur. At present the discovery of a clean-cut new species is a rare event, and discovery of truly new genera has virtually ceased. We may be sure that we now know nearly, if not quite, all the specific and higher taxa of those classes. Collecting of recent insects, on the contrary, still produces many new species and a proportion of higher taxa, so that knowledge of those animals is evidently still highly incomplete. Discovery of new taxa among fossils has not ceased in any group and is still accelerating in most. It has, nevertheless, definitely slowed down for some groups, such as the ammonites, and it appears that our knowledge of those groups (as far as they are actually accessible in the rocks) is nearing completion.

It is, further, possible to estimate within broad limits how large an ancient biota is likely to have been and therefore what percentage of its taxa are represented in known collections. One approach is exemplified by the unusual but enlightening example of the Pleistocene mammals of Florida.

There is reason to believe that all or nearly all the native species of mammals living in peninsular Florida in historic times were already present there in the Late Pleistocene. The time elapsed is not long enough for the normal evolution of distinctly new species. Florida's geographic and climatic situation is such that the immigration of species in Late Pleistocene and Recent is improbable. (In many areas it is not only probable but has in fact occurred to a significant extent.) On this reasonable postulate, the percentage of Recent species recovered from Late Pleistocene deposits should reflect, in some degree, the percentage of recovery of the whole Pleistocene fauna. The conditions are favorable, because Late Pleistocene deposits rich in fossil mammals are widespread in peninsular Florida and have been intensively studied, even though not, as yet, by methods likely to recover all extremely small or exceptionally rare forms.

The results are given in Table 1-5. As would be expected a priori and from experience elsewhere, recovery of very small and of arboreal mammals is poor—one-third to one-half of species presumed present—and recovery of volant mammals (bats) is altogether inadequate—less than one-tenth. On the other hand, recovery of other species (terrestrial, fossorial, or amphibious and of moderate to large size) is almost complete, only one recent species lacking out of twenty-three. (The missing species is the mink, *Mustela vison*, which was probably rare; the Pleistocene record of the cougar, *Felis concolor*, is not quite certain but is probable.) Recovery is 25 per cent for very small, arboreal, and volant mammals combined, 96 per cent for others, and 63 per cent overall.

These figures do not at once yield a good estimate of recovery of all Pleistocene species, because extinction was differential with respect to chances of recovery. More than half (56 per cent) of the known Pleistocene species are either totally extinct (48 per

TABLE 1-5

*Recovery of Recent Species of Mammals from Pleistocene of
Peninsular Florida and Estimation of Total Pleistocene
Species and Recovery*

(New Compilation of Data from Many Sources)

(a) Recent Species and Their Recovery from the Pleistocene

		Total Recent Species		Recent Species Known from Pleistocene		Recovery of Recent Species
		(No.)	(Per Cent)	(No.)	(Per Cent)	(Per Cent)
(1)	Low-recovery groups					
	(a) Head–body length					
	<4 in	6	14	3	11	50
	(b) Arboreal	3	7	1	4	33
	(c) Volant	11	26	1	4	9
	Total	20	47	5	19	25
(2)	High-recovery groups Head–body length <4 in, terrestrial, fossorial, or amphibious	23	53	22	81	96
	Grand total	43	100	27	100	63

(b) Pleistocene Fauna

		Also Recent in Florida	Also Recent Not in Florida	Extinct	Total	Per Cent
(1)	Low-recovery groups					
	(a) Head–body length					
	<4 in	3	0	1	4	7
	(b) Arboreal	1	0	0	1	2
	(c) Volant	1	0	1	2	3
	Total	5	0	2	7	11
(2)	High-recovery groups	22	5	27	54	89
	Grand totals	27	5	29	61	100
	Per Cent of Pleistocene Total	44	8	48	100	—

(c) Estimation of Total Pleistocene Fauna
from Numbers of Known Species

		Known No.	Postulated Recovery (Per Cent)	Estimated		Postulated Recovery (Per Cent)	Estimated	
				(No.)	(Per Cent)		(No.)	(Per Cent)
(1)	Low-recovery groups							
	(a) Head–body length <4 in	4	50	8	9	—	—	—
	(b) Arboreal	1	33	3	3	—	—	—
	(c) Volant	2	9	22	25	—	—	—
	Total	7	21*	33	37	25†	28	33
(2)	High-recovery groups	54	96	56	63	96	56	67
	Grand totals	61	69*	89	100	73*	84	100

*Weighted mean recovery; figure not in previous tables.
†Unweighted mean recovery for total low-recovery groups.

cent) or no longer live in Florida (8 per cent) (see Table 1–5b). Of these, only a minute proportion (6 per cent of species now extinct or otherwise absent in Florida) belong in the low-recovery categories of very small, arboreal, or volant species. That figure is of course biased by the very fact of low recovery. If we use the estimates of recovery derived from living species(Table1–5c)the low-recovery groups probably constituted from 33 to 37 per cent of the whole Late Pleistocene fauna. In the Recent fauna, 47 per cent of species belong to low-recovery groups. Thus high-recovery groups (mostly large terrestrial species) were more numerous in the Pleistocene, and a larger percentage of them became extinct, either locally or completely. On this basis, overall recovery from the Pleistocene is estimated as about 70 per cent or more (69 to 73 per cent on different postulates), whereas total Pleistocene recoveries of Recent species are only 63 per cent. Both figures will, of course, be increased by more intensive collecting, especially if this is more directed toward recovery of the smallest and rarest forms.

That method is inapplicable to those fossil faunas—the great majority—in which the presence of Recent taxa is not an acceptable postulate. It still is possible to judge broadly whether the faunal composition and diversity indicated by the samples is representative of a fauna that is reasonable on various, especially ecological, grounds. In most faunas there will, for instance, be a reasonably expectable ratio between herbivores and car- nivores, small and large animals, etc. One example among the many possible is given in Table 1–6. The comparison of the samples of Middle Oligocene nonvolant mammals of western South Dakota with Recent fauna of the same region indicates that the ancient fauna must have been richer. The sampling cannot be complete but already includes more genera than the Recent fauna. The composition also is evidently different, the ancient fauna much richer in ungulates and poorer in rodents, a difference doubtless exaggerated by, but not wholly due to, sampling error. On a balance of evidence, the

TABLE 1–6

Comparison of Genera in an Ancient and Two Recent Faunas of
Non-volant Land Mammals*

| | Western South Dakota | | | | | |
| | Middle Oligocene | | Recent | | Recent Portuguese East Africa | |
	(No.)	(Per Cent)	(No.)	(Per Cent)	(No.)	(Per Cent)
Insectivores and small to moderate om- nivores	3	7	3	7	12	15
Small herbivores (mostly rodents)	13	30	22	54	25	32
Large herbivores and a few omnivores (mostly ungu- lates)	19	44	5	12	24	30
Carnivores	8	19	11	27	18	23
Total	43	—	41	—	79	—

* Data for the fossil fauna somewhat modified from Scott and Jepsen (1936–41); data for Portuguese East Africa mainly from Ellerman, Morrison–Scott, and Hayman (1955); data for Recent South Dakota somewhat modified from Over and Churchill (1941).

environments differed in that this area in the Middle Oligocene was lower, wetter, with more equable temperatures, and more extensively forested. The environment of Portuguese East Africa is surely not the same but is probably more like Middle Oligocene South Dakota than is Recent South Dakota. The general ecological makeup of the mid-Oligocene and East African faunas is, indeed, more similar, despite the fact that the actual taxa are completely different below the family level. These comparisons and some other considerations suggest that surely half and probably two-thirds or more of the Middle Oligocene genera are known and that those not yet known are mainly carnivores (individually much less abundant than herbivores) and very small mammals (with less recoverability than large mammals by previous collecting methods.)

A final point about the adequacy of the record to be made here (among many others that might be made) has to do mainly with the relationships between bias and the current aims of historical studies. The list of sources of bias given on a previous page is downright appalling if viewed in isolation. It might seem to vitiate any paleontological generalizations. Yet the mere fact that these biases are known and listed makes reliable generalizations possible. It makes it increasingly possible to seek out samples least biased with respect to any particular generalization. It is also becoming increasingly possible to make suitable allowance for biases that are present in otherwise usable samples.

In Darwin's day and down practically to our own, concern with the adequacy of the fossil record was almost entirely with its degree of completeness, taxonomically and anatomically. We are, of course, still interested—even primarily interested—in that point, but we are also increasingly interested in others. Many of the questions of bias would hardly arise, were it not for the fact that we are now demanding much more and different kinds of information from the record. We are, for instance, investigating the relative abundance of individuals of various species within a fauna or its ecological makeup (summary example in Table 1-6; see also, e.g. Shotwell, 1955). It is even becoming possible to construct mortality curves and age pyramids or to measure selection intensity in fossil populations (e.g. Kurtén, 1953, 1958). Such studies as these greatly increase the usefulness and significance of the record, but they bring up questions of sampling bias that were of no importance when the only concern was with taxonomy and anatomy. For some problems—for instance, those of geographic subspeciation—bias in available samples is so great that paleontologists have not yet successfully coped with them. For an increasing range of problems, however, conscious attention to bias in selecting samples and devising methods permits exemplary generalizations.

[Considerations of bias and sampling have a bearing on the fact that] it is a feature of the known fossil record that most taxa appear abruptly. They are not, as a rule, led up to by a sequence of almost imperceptibly changing forerunners such as Darwin believed should be usual in evolution. A great many sequences of two or a few temporally intergrading species are known, but even at this level most species appear without known *immediate* ancestors, and really long, perfectly complete sequences of numerous species are exceedingly rare. Sequences of genera, immediately successive or nearly so at that level (not necessarily represented by the exact populations involved in the transition from one genus to the next), are more common and may be longer than known sequences of species. But the appearance of a new genus in the record is usually more abrupt than the appearance of a new species: the gaps involved are generally larger, that is, when a new genus appears in the record it is usually well separated morphologically from the most nearly similar other known genera. This phenomenon becomes more universal and more intense as the hierarchy of categories is ascended. Gaps among known species are sporadic and often small. Gaps among known orders, classes, and phyla are systematic and almost always large.

These peculiarities of the record pose one of the most important theoretical problems in the whole history of life: is the sudden appearance of higher categories a phenomenon of evolution or of the record only, due to sampling bias and other inadequacies? I have discussed this question at such great length elsewhere (especially Simpson, 1953) that I need do no more here than summarize with irreducible brevity and add comments on a few points that I may not have sufficiently stressed before.

In the first place, the "either . . . or" question is not necessarily answered exclusively by either one of the alternatives. There is little doubt that the rise of a higher category has frequently involved exceptionally fast evolution, change at rates considerably higher than those normally involved in the better-recorded sequences. To that extent the sudden appearance of higher categories is almost certainly influenced by a real evolutionary phenomenon. But this does not exclude the possibility that the origins of those categories were by graded transition, even though some phases occurred at unusually rapid rates. The gaps, then, would not, as such, truly represent an evolutionary phenomenon ("saltation," "typostrophism," "magnimutation," etc.) but only a sampling effect.

Second, there are well-grounded theoretical reasons why organisms in the major transition sequences may be less recoverable as fossils, so that the record would be biased against representation of them in our collections. The rapidity of their evolution would, in itself, produce such a bias. In such transitions the populations involved were probably in most cases, if not in all, notably smaller in size and much more restricted in distribution than their most successful, abundant, and widespread descendants that appear with apparent suddenness in the record as we know it. Those are potentially very strong biasing factors. Moreover, many higher categories appear in the record at times (e.g. Triassic, Paleocene) following geographic conditions such that there were few deposits of facies appropriate for preservation of the ancestors of the groups that do appear suddenly. There are other, but probably less important, biasing factors (Simpson, 1953). In sum, these considerations strongly suggest that the probability of collecting a representative of a species in a transition to a new higher category is decidedly less than for species within the category after it is fully established.

There is a further point that has been noticed before but that seems to me now to have been insufficiently emphasized or appreciated. Even if sampling were quite unbiased, with any one species just as likely to be preserved and found as any other, incomplete sampling and taxonomic conventions would produce, purely as artifacts, gaps between taxa that would be more systematically present and larger as the hierarchy was ascended. This can be demonstrated not only as a logical statistical consequence but also by simple paper experiments, one of which is summarized in Table 1–7. Draw a phylogenetic tree, either hypothetical or deduced for a real group of organisms. Divide all its lineages into numbered species, postulated as of the same "size" (taking up equal lengths on the lines). Draw a sample of say 10 to 15 per cent of the species at random and mark these as "known," i.e., discovered as fossils, the others being "unknown." The resemblance and affinity between any two species is postulated as approximately proportional to the least number of intervening species along lineages in either direction. Classify the "known" species into genera, families, etc. as you would actual specimens.

Most of the "known" species will be isolated, but a few sequences of two, three, or

TABLE 1–7

*Results of Experiment in Sampling a Phylogeny**
(Experiment Described in Text)

Category	No. "Known" Taxa	Species) between Gaps (Number of "Unknown" Taxa within Single Units of Next Higher Category		
		(No.)	(Range)	(Mean)
Species	35	26	0–8	2.8
Genera	16	12	2–13	8.2
Families	4	3	8–18	10.3

* Number of original species presumed to have existed: 350; number selected by random sampling: 35.

possibly more will probably occur. In the experiment of Table 1–7 there were three sequences of two contiguous species and one of three. There will also be longer sequences with a few short gaps. Tabulate the gaps by length (number of intervening "unknown" species) for species within genera, genera within families, families within superfamilies, etc. It will be found that the mean length of gaps is small for species, larger for genera, still larger for families, and so on. That is just the situation in the record as actually known. The lower limit of the range for any category will be not more and generally distinctly less than for the next higher category. For species it will probably be zero. For genera it may be zero, but this is unlikely. For still higher categories it will almost certainly be above zero: there are systematic gaps between those categories, as in the actual record. The upper limit does not decrease and tends to increase, the higher the category. (In Table 1–7, based on a comparatively simple phylogeny that tends to minimize the contrasts, the upper limit is the same for genera and families.) Similarly, in the known record the largest gaps are between the highest categories.

Thus the observed gaps in the record are fairly analogous to what would be expected from random sampling of originally continuous sequences without gaps. The correspondence becomes exact if allowance is made for biases almost certainly present. The presence, relative extent, and distribution of the gaps can be *predicted* from the postulate that no gaps were really present in the original structural and phylogenetic sequences. Other predictions can be made, and their fulfilment seems further to establish the postulate as true beyond serious question. Here are some examples:

(1) On this postulate, intensive sampling should rarely, but occasionally, produce forms scattered more or less randomly within what are otherwise major gaps. This happens: *Archaeopteryx* in the reptile–bird gap is the most famous example.

(2) Direct generic sequences should be more common than direct specific sequences. That is true of the record, and it may appear anomalous in view of the fact that gaps between genera are likely to be more frequent and large than between species. The prediction follows from these considerations: sampling strongly favors the largest, most widespread specific populations. But, on an average, smaller and more local populations are more numerous. A large population is no more likely and may well be less likely to give rise to a new genus than a small population. Therefore, the chance of finding successive genera represented by species that are not necessarily successive is greater, on an average, than the chances of finding successive species.

The prediction would indeed still follow if any one species of a genus were as likely as any other to give origin to a new genus. The basically sufficient conditions for the prediction are that genera have, on an average, more than one species, which is obviously true, and that two successive species in one genus do not usually both have much higher probability of discovery than the average for all the species of known genera, which must hold more often than not.

(3) Some taxa should reappear after absence from the record for spans comparable to the gaps in the record characteristic of the hierarchic categories involved. That is notoriously true to the living coelacanths and monoplacophoran mollusks, the youngest known fossils of which are Cretaceous and Silurian, respectively. Other, less sensational examples are fairly common and often tend, as would also be predicted, to be inversely proportional to the knowledge (or intensity of sampling) of the group involved. For instance, for scyphozoans, a poorly sampled group, there is an enormous gap from Cambrian to Jurassic and for condylarths, a well-sampled group, a much shorter gap from Eocene to Miocene.

Source of this partial and modified reprinting:

(1959) *Evolution after Darwin* (Edited by SOL TAX) University of Chicago Press, Vol. I, pp. 117–180.

References

CAILLEUX, A. (1954) "How Many Species?" *Evolution*, VIII, 82–83.
ELLERMAN, J.R., MORRISON-SCOTT, T.C.S., and HAYMAN, R.W. (1953) *Southern African Mammals 1758 to 1951: A Reclassification*, London: British Museum.
KURTÉN, B. (1953) "On the Variation and Population Dynamics of Fossil and Recent Mammal Populations," *Acta Zool. Fennica*, No. 76.
KURTÉN, B. (1958) "Life and Death of the Pleistocene Cave Bear," *ibid.*, No. 95, pp. 1–59.
LAURIE, E.M.O., and HILL, J.E. (1954) *List of Land Mammals of New Guinea, Celebes, and Adjacent Islands*, London: British Museum.
MATTHEW, W.D. (1937) "Paleocene Faunas of the San Juan Basin, New Mexico," *Trans. Amer. Phil. Soc.*, N.S., XXX, i–viii, 1–510.
MULLER, S.W., and CAMPBELL, A. (1954) "The Relative Number of Living and Fossil Species of Animals," *Systematic Zoöl.*, III, 168-70.
NEWELL, N.D. (1959) "The Nature of the Fossil Record," *Proc. Amer. Phil. Soc.*, CII, 264–285.
OVER, W.H., and CHURCHILL, E.P. (1941) "Mammals of South Dakota," Vermillion (S.D.) (Mimeographed by the museum and zoölogy department of the University of South Dakota.)
SCOTT, W.B., and JEPSEN, G.L. (1936–41) "The Mammalian Faunas of the White River Oligocene," *Trans. Amer. Phil. Soc.*, N.S., Vol. XXVIII, Parts I–V. [A.E. WOOD is author of two parts.]
SHOTWELL, J.A. (1955) "An Approach to the Paleoecology of Mammals," *Ecology*, XXXVI, 327–37.
SIMPSON, G.G. (1937) The Fort Union of the Crazy Mountain Field, Montana, and Its Mammalian Faunas, *U.S. Nat. Mus., Bull.* 169.
SIMPSON, G.G. (1952) "How Many Species?" *Evolution*, VI, 342.
SIMPSON, G.G. (1953) *Major Features of Evolution*, New York.
TEICHERT, C. 1956. "How Many Fossil Species?" *Jour. Paleontol.*, XXX, 967–69.

Continuity and Discontinuity in the Fossil Record

Comment

This brief discussion expresses further, later thoughts about the adequacy of the fossil record. It is here reprinted, slightly modified but almost in full, from an invited "editorial" in a specialized journal devoted to research on the relatively fossil-poor Precambrian part of the history of life. It closes with an almost too moderate judgment of a view about the pattern of evolution that became widely noticed in 1978 and into 1979 and that seems to me to have been of relatively minor novelty or importance. I have here added a reference to its first clear expression, omitted in the original publication of this paper.

The Ediacara fauna, mentioned in the second paragraph, is named for a locality in South Australia where it was found. Similar faunas have since been found also in Europe and Africa.

Text

This gloss on Precambrian paleontology was written with the idea that experience with the more richly documented Phanerozoic might suggest principles applicable to Precambrian studies.

The most striking feature of the Precambrian fossil record is its appalling incompleteness. That great span, covering at least four-fifths of the existence of organisms on this planet, has a known record characterized by vastly more lacunae than actual specimens of such beings. To the best of my knowledge animals, strictly speaking, as unmistakable metazoans, are not known before the Ediacara fauna, which occurs so late that it has been suggested that it be referred to the earliest Cambrian rather than to the latest Precambrian. Yet the animals of the Ediacara fauna are so varied and some of them are so complex that metazoans must have existed, without leaving traces now known, for a very long earlier time.

It has often been said that a continuous known fossil record starts with the Cambrian, hence the term "Phanerozoic" from the Greek *phaneros*, "visible" or "manifest". Yet to consider the Phanerozoic fossil record as continuous is quite misleading. It is true that fossils of some kind are known from some place at geologically short or virtually infinitesimal intervals from the early Cambrian to the Recent. Yet we all know that for any one taxon and for any area over a considerable span of geological time the record is virtually never continuous.

The fossil record of mammals, the group of animals with which I am most familiar, provides an instructive small-scale model of the large-scale contrast between the Precambrian and the Phanerozoic records of organisms in general. The known record of unmistakable mammals begins in the Late Triassic, approximately 195 million years before present and approximately 130 million years before the beginning of the Cenozoic. Considering this as a scaled model in the present connection, those 130 million years in the Mesozoic are analogous to the Precambrian and the approximately 65 million years of the Cenozoic are analogous to the Phanerozoic. Thus at least two-thirds of the history of mammals had occurred before the Cenozoic, "Age of Mammals," began. That is comparable to the fact that at least four-fifths of the history of life had elapsed before the Phanerozoic began.

The point here is that for the span of at least 130 million years when we know that mammals existed there is no record of them at any place on earth for about 100 million of those years. For Australia and Antarctica there is no known record of Mesozoic mammals at all. In Africa the only known record for about 125 of the 130 million years of existence of Mesozoic mammals is a single jaw, without preserved teeth. In South America there was no known record of surely identifiable Mesozoic mammals until a few scraps were recently found in rocks of the very end of that era. From Europe just one mammal tooth is from the last 75 million years or so of the Mesozoic. The records are somewhat better for Asia and North America, but even for those continents there are almost incomparably more vacant gaps than positive knowledge.

In the Cenozoic the known record of mammals is continuous in the same sense as the continuity of known record for animals in general during the Phanerozoic: some kinds of mammals at some places are known for effectively all ages from early Paleocene to Recent. Nevertheless, records of particular taxa or in particular regions are almost always incomplete. It is a generalization as true for mammals in the Cenozoic as in the Mesozoic and as true for animals in general in the Phanerozoic as in the Precambrian that taxa at all hierarchic levels from species to kingdoms usually appear abruptly in the known fossil record without ancestral taxa known at the same or on a lower hierarchic level. In spite of the fact that literally millions of fossils have been collected since 1859, it is as true today as it was when Darwin wrote *The Origin of Species* that the fossil record is woefully incomplete.

To other paleontologists, with a few exceptions, I may seem to belabor the obvious. Worse yet, I may seem to give a counsel of despair. Neither is true. The gaps in the fossil record themselves require emphasis and study. The record as it stands is, nevertheless, adequate for certain purposes.

In pre-Darwinian days the abrupt appearances of the new taxa, then already known and even more impressive than now, was commonly interpreted as indicating that each taxon had been divinely created at about the time of its appearance in the sequence. An alternative less common but supported by Cuvier, the greatest paleontologist of his time, was that all taxa had been divinely created once and for all but that by migration from elsewhere they appeared suddenly in Europe, for which alone there was then a fairly good known succession.

Those views are not now held by any competent paleontologist. Yet there are paleontologists who hold that the known record is now near enough to completion to be taken at its face value. The late Otto Schindewolf maintained that the origin of new taxa occurred instantaneously by mutations with large effects usually resulting from changes in early ontogenetic stages. Another recent view, called "punctuated equilibrium" by Gould and Eldredge, is that new taxa usually arise in small marginal populations (hardly ever known as fossils) by evolution so rapid as to appear instantaneous in terms of detectable geological time.

The premise that the known record can be taken at anything like its face value is untenable. The Crossopterygii, usually ranked as an order, have a known fossil record, with gaps of course, from the early Devonian to the late Cretaceous. Then none are known until the Recent, when there is a living crossopterygian, *Latimeria*. Obviously there were unrecorded crossopterygians during the blank 65 million years of the Cenozoic, and they underwent marked evolution during that time because the now living form is unlike any of the known fossils at the family level if not higher. That is a single example, striking and widely publicized but not isolated or really exceptional. More often than not there are blanks in the record at times when, and even where, we know that certain taxa were in existence.

That is a positive conclusion drawn from the fossil record itself and bearing on the negative aspects of the known record. The point here is that the fossil record does not now and almost certainly never will provide in a truly complete and definitive way, a history of life, a phylogeny of any major taxa, or a comprehensive theoretical interpretation of evolution. What it does provide is a body of restraints relevant to all three of those subjects and several others. The restraints become more numerous and more imperative as knowledge of the fossil record increases. For questions of history and phylogeny of mammals the restraints are stronger and tighter in the Cenozoic than in the Mesozoic. For such questions about organisms in general they are stronger and tighter in the Phanerozoic than in the Precambrian.

Only for relatively minor problems are the restraints absolute. It has long been known that rates of evolution vary, that they tend to be higher when new taxa are originating, and that they must have been markedly higher when major taxa, at the level, say, of phyla or classes, were arising than for taxa at lower hierarchic levels. That is a deduction from the record especially important for Precambrian research. The restraints do not suffice to refute the hypothesis that in the origin of clearly distinct taxa the rates are always too rapid to be measured in geological time. They do not yet suffice to rule out the possibility that that is sometimes or perhaps usually the case, which seems to me to be the only new element in the current discussion of "punctuated equilibrium."

Source of this slightly modified reprinting

(1978) Editorial Essay, *Precambrian Research*, Vol. 7, pp. 101–103.

Other relevant references

GOULD, S.J., and N. ELDREDGE (1977) Punctuated Equilibria: the Tempo and Mode of Evolution Reconsidered, *Paleobiology*, Vol. 3, pp. 116–151.
SCHOPF, J.W. (1975) Precambrian Paleobiology, Problems and Perspectives, *Ann. Rev. Earth and Planetary Sci., Vol. 3, pp. 213–249.*

History of a Fauna

Comment

The following text is an example on a fairly large scale of some methods of interpretation and representation of evolutionary history. It involves a considerable group of animals, the Class Mammalia, in a large area, the zoological region called Nearctica, which is geographically nontropical North America, through a fairly long stretch of geological time, the approximately 65 million years of the Cenozoic era. The outcome of that history, the living mammalian fauna of this region, is analyzed as the result of the rather complex faunal changes since the beginning of the era, which is so dominated on land by the mammals that it is often called the Age of Mammals although it comprises only about a third of the time in which mammals have existed as such.

The parts of the North American fauna of land mammals not considered here are those of the present Neotropical zoological region, which geographically is essentially Central America (including the tropical parts of Mexico) and all of South America. I have summarized the history in that region in a recent book, *Splendid Isolation*.

The following text is based on part of the Condon Lectures delivered in January, 1953, at the University of Oregon, Oregon State College (now University), and Portland State College (also now University) and printed later that year by the Oregon State System of Higher Education. I have here made a number of minor changes in the text, mostly to delete or explain references to other parts of the lectures not included in the present work.

Discoveries and changes in classification since 1953 have modified the data represented in the figures, here retained without change. The general character of these is still not markedly different, and in any case they stand as examples of analyses of faunal data and methods of their presentation. Other examples of faunal history are given in Chapter 3 of the present work.

Text

The richness and general nature of the present North American fauna of land mammals are shown in Table 1–8. The sequence of earlier faunas throughout the Cenozoic is better known for this than for any other continent. The broadest feature is great increase in fundamental diversity, the number of major groups or orders, through the Paleocene and into the Eocene. After the early Eocene the number declined markedly into the middle Oligocene. Since the middle Oligocene the orders present have not always been the same, but their number has remained nearly constant (Fig. 1–1). There have been continuous, striking changes in composition of the fauna through the Cenozoic, as shown in Fig. 1–2.

The most radical turnover, mostly by introduction of new groups and extinction of older ones, was during the Paleocene and Eocene (Fig. 1–3). A few early Paleocene orders were still present in the late Eocene, but those survivors constituted a decidedly minor part of the fauna except for the Order Carnivora. Even within that order the turnover was great, for most late Eocene carnivores were of quite different groups from their early Paleocene forerunners. Since the late Eocene there have been profound changes, to be sure, but these have been far less radical than the Paleocene–Eocene changes. As regards the broad orders, the most noteworthy changes since the late Eocene have been the regional extinction of the Perissodactyla (horses, rhinoceroses, tapirs, and their extinct relatives) and continuous great expansion of the rodents, which now constitute about half of the North American fauna (in terms of both genera and species).

The accompanying figures (1–2, 1–3) show changes in percentage composition. It is more difficult to estimate changes in absolute numbers of genera and species, because we do not know all those groups before the recent and it is impossible to be sure what

TABLE 1–8

Synopsis of Recent Nearctic Land Animals

Orders are named in capitals and under each are listed the included families. Genera and species counted are nonaquatic, native, mammals, exclusive of bats, living in the United States and Canada. A few species that barely or occasionally cross the Mexican border but are of mainly Neotropical groups are omitted.

Orders and families	No. of genera	No. of species	Orders and families	No. of genera	No. of species
MARSUPIALIA	1	1	RODENTIA, (cont.)		
Didelphidae,			Heteromyidae,		
opossums	1	1	pocket mice and		
INSECTIVORA	10	36	kangaroo rats	3	36
Soricidae,			Castoridae, beavers	1	1
shrews	5	29	Cricetidae, native		
Talpidae,			mice and rats	16	63
moles	5	7	Zapodidae, jumping		
EDENTATA	1	1	mice	2	3
Dasypodidae,			Erethizontidae, por-		
armadillos	1	1	cupines	1	1
LAGOMORPHA	3	14	CARNIVORA	16	30
Ochotonidae,			Canidae, wolves		
pikas	1	1	and foxes	4	7
Leporidae,			Ursidae, bears	1	4
rabbits	2	13	Procyonidae, raccoons, etc.	2	2
RODENTIA	35	165	Mustelidae, weasels,		
Aplodontidae,			skunks, etc.	8	14
mountain beaver	1	1	Felidae, cats	1	3
Sciuridae,			ARTIODACTYLA	5	6
squirrels, etc.	8	51	Antilocapridae,		
Geomyidae,			pronghorns	1	1
pocket gophers	3	9	Bovidae, bison, sheep		
			goats	4	5

TOTALS: Orders, 7; Families, 21; Genera, 71; Species, 313.

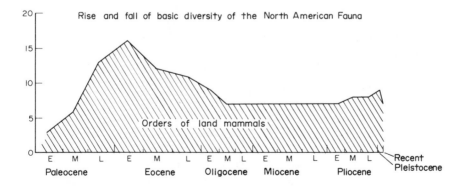

FIG. 1–1. Changes in fundamental diversity of mammals during the Cenozoic era in North America in terms of numbers of orders of land (nonflying and nonaquatic) mammals. The epochs of the Cenozoic are named in sequence, oldest at left, and all except the last two, relatively short epochs are divided into early (E), middle (M), and late (L) parts. The epochs are approximately scaled to their relative lengths in years.

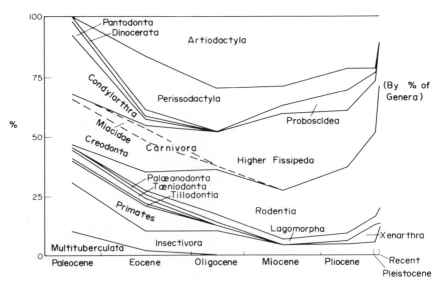

FIG. 1–2. Changes in known composition of the faunas of land mammals in North America during the Age of Mammals (Cenozoic) in terms of percentage of genera referred to each order. The figures for the orders are arbitrarily graphed at the middle of each epoch. [The exact figures have been changed by later discoveries, but the pattern remains essentially the same. The genera here plotted as Creodonta are now often referred to orders other than the Carnivora. Note added in 1979.]

proportion of them we do know for various times. The general impression is that the numbers of genera and species increased greatly and rather steadily from the early Paleocene onward perhaps into the Miocene. In the Pliocene and Pleistocene the numbers were still high but perhaps lower than in the Miocene, and since some time in the Pleistocene they have almost certainly decreased.

Several North American families, Aplodontidae (sewellels), Geomyidae (pocket gophers), Heteromyidae (Kangaroo rats), and Antilocapridae (pronghorn "antelopes"), many genera, and almost all species are now almost completely confined to Nearctica (are mainly endemic there). However, most of the families also occur on other continents, and so do more than half of the genera and even a few species. (For example, our wolf, *Canis lupus*, is not specifically separable from that of Eurasia, nor is our puma, *Felis concolor*, from that of South America.) Other factors are also involved, but these varying degrees of distinctiveness of members of our fauna are in part a reflection of the fact that the fauna is complexly stratified. The complexity is, indeed, so great that all the separate strata have not yet been clearly recognized or their members sorted out. It is, nevertheless, possible to make an approximation of broad stratal analysis by determining the ages of the various families of mammals present now and through the Cenozoic.

Fig. 1–4 presents data for first appearance and survival of known families of land mammals from Late Cretaceous onward and for the whole world. Offhand one might expect that first appearances would decrease and survival would increase as time went on. There is perhaps some such tendency, but it is not strong or regular. First appear-

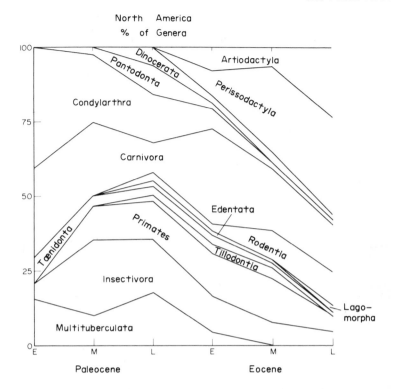

North America
% of Genera

FIG. 1–3. Changing composition of the North American fauna of land mammals during the first two epochs of the Cenozoic. As in Fig. 1–2 the graph is based on percentages of genera referred to each order, but here given the early, middle, and late parts of each of the two epochs and placed at the midpoints of those parts. The striking feature is the radical turnover in basic composition of the faunas. [Although later work has changed details, the general pattern is still the same. This is also true for Figs. 1–4, 1–5, and 1–7. Note added in 1979.]

ances were highest in the Eocene and survival is greatest from the Oligocene. Fig. 1–5 presents in different graphic style a similar analysis for appearance and survival in North America, only. Effects of the great Paleocene–Eocene turnover (Fig. 1–3) are evident: in North America no Paleocene family survived later than the Miocene. As for the world as a whole, in North America more families of Oligocene first appearance still survive than for any other epoch. The stratal complexity is reflected by the fact that the families of mammals now living in North America first appeared here at various times from well before the end of the Cretaceous (at least 80 million years ago) onward.

When a new group of land mammals, say a family or a genus, appeared in North America it came from one or another of the following sources:

North America, itself, by evolutionary change from an older group;
South America;
Asia;
Europe.

Other possibilities, perhaps such as spread from the West Indies, Africa, or the Southwest Pacific, perhaps cannot wholly be ruled out, but they are highly improbable

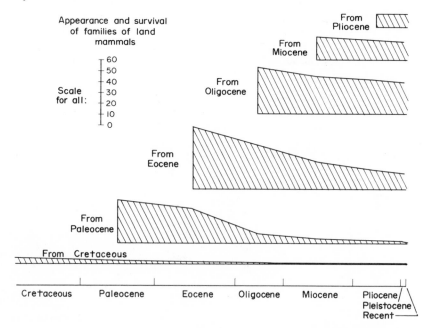

FIG. 1–4. Appearance and survival of families of known land mammals for the world as a whole. The number of families is shown to scale (vertical height) at the middle of each epoch in which they are first known and thereafter the number of those remaining (nonextinct) is scaled at each epochal midpoint.

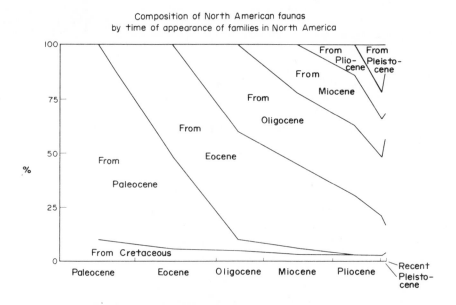

FIG. 1–5. The changing composition of North American families of land mammals in terms of the age of the families from their first appearance in North America.

and there seems to be no evidence at all that any North American land animals ever did come directly from any geographic source other than the four just named.

The occurrence of truly related animals in both of two regions is conclusive evidence that there has been some degree of faunal interchange between the regions at some time in the past. It does not necessarily indicate a direct interchange. For example tapirs now occur only in the Oriental and Neotropical Regions, but there has certainly never been faunal interchange directly between those two regions. The distribution of tapirs is the result of interchanges between Oriental and Palearctic, Palearctic and Nearctic, and Nearctic and Neotropical Regions, as their occurrence as fossils in the Palearctic and Nearctic plainly demonstrates. (The Oriental Region comprises southern Asia and adjacent islands. The Palearctic includes central and northern Asia, Europe, and northern Africa.)

As a rule, the more closely related are animals in two different regions the more direct and more recent was a faunal interchange or continuity between those regions. In attempting this sort of analysis between the faunas of different continents it is usually of little practical use to seek extremely closely related groups, those classified as belonging to the same species. The processes of spread from one continent to another and of separate evolution on a different continent are usually accompanied by evolutionary changes that cause the animals to be considered different species. Often, however, they are still recognizable as belonging to the same genus. It is, then, a justified postulate of study of continental faunal interchanges that the occurrence of the same genus on two different continents indicates distribution over both not long before, geologically speaking. "Not long before" is a vague expression and the time doubtless varies greatly, possibly up to several million years, but it is likely to be short in proportion to the 65 million years, more or less, of the Age of Mammals.

Each occurrence of related animals on two continents is an item of evidence for intermigration between them (Fig. 1–6). By study of faunas at successive times during the Cenozoic, such items can be added together for each definable time span. A picture is then obtained of increase and decrease in migration intensity.

(a)   (b)

Fig. 1–6. A characteristic item of evidence for former faunal connection between what are now two separate continents. These are upper cheek teeth (the left premolars and molars in crown view) of small weasel-like extinct mammals. Although somewhat different in detail they are clearly closely related and are referred to the same genus, *Palaeogale*. Specimen (a) is from Mongolia in Asia, (b) from Colorado in North America.

Faunal interchange between Eurasia and North America has been frequent and sometimes intense. Some of the items of evidence are summed up in the graph of Fig. 1–7. Times of exceptionally intense intermigration included early Eocene, early Oligocene, late Miocene, and Pleistocene. Middle (perhaps also early) Miocene, and middle and late Pliocene were times of less but still significant intermigration. In the middle Eocene, middle and late Oligocene, and early Pliocene there were times of little or no intermigration, and this is true also of the Recent, when we know that strong barriers exist.

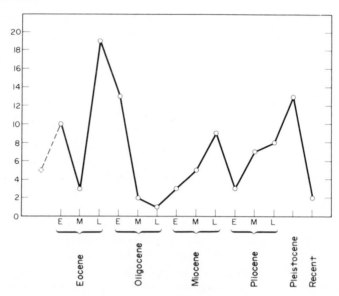

FIG. 1–7. Numbers of items of evidence for faunal connection between Eurasia and North America during most of the Cenozoic. The ups and downs of the graph are somewhat roughly indicative of the degree of continental isolation at the times shown. Thus there was relatively little isolation in the early Eocene, late Eocene to early Oligocene, late Miocene, and middle Pliocene to Pleistocene, and relatively strong isolation in the middle Eocene, middle Oligocene to middle Miocene, early Pliocene, and Recent. [It has now become fairly clear that the main connection of North America to Eurasia in the early Eocene was with Europe but that those thereafter were with Asia. Note added in 1979.]

(There are, to be sure, some items of evidence for intermigration at the stated times but these evidently reflect either slightly earlier migrations or the fact that interruption may not have continued through the whole time span in question.)

Such evidence permits further inferences. When interchange was moderate to high, there certainly was a land route between Eurasia and North America, because considerable essentially simultaneous spread in both directions can hardly be accounted for in any other way. It further follows that the land connection was probably interrupted by a sea barrier when interchange was slight or nil, although this negative inference is less certain than the positive inference of presence of a land bridge at other times. Even the most intense interchanges involved only a fraction of the whole faunas of the continents. It may therefore be concluded that the connection was a filter bridge rather than a corridor. Study of what kinds of mammals were passed or stopped by the filter suggests (although the evidence is not yet completely conclusive) that those stopped were the ones adapted exclusively to tropical climates or, at least, not at home in the cooler climates of each epoch. Studies of resemblances between North American faunas and those of different parts of Eurasia further demonstrate that during and after the late Eocene, at least, resemblance is closest with faunas of northeastern Asia. All this and other evidence supports the conclusion that the filter bridge was in the north and that it connected Asia and North America more or less in the regions now designated as eastern Siberia and Alaska. There was also an early Cenozoic land connection directly with Europe.

Source

(1953) *Evolution and Geography: An Essay on Historical Biogeography with Special Reference to Mammals,* Condon Lectures, Oregon State System of Higher Education. (The part here reproduced in slightly modified version is on pages 43–50 of the printed pamphlet.)

Periodicity in Vertebrate Evolution

Comment

The fossil record bears on many different evolutionary problems and hypotheses. The following text is one example of these and includes a method for studying and illustrating data relevant to it. Students of the history of life early noted that there have been times when evolutionary change seems to have been particularly active and widespread. Students of the history of the physical earth noted that there have also been times when activity in the earth's crust, diastrophism, seems to have been particularly active and widespread. It was believed by many that there was a regular, rhythmic periodicity in both these respects. A catchy expression of this idea was provided by the notable Dutch geologist J.H.F. Umbgrove in 1942 when he titled a widely acclaimed book *The Pulse of the Earth.*

A further idea was that the biological episodes, picturesquely but misleadingly called "evolutionary explosions," had coincided with the geological episodes, also picturesquely and also misleadingly called "tectonic revolutions." An almost inescapable corollary was that there might be a causal connection between the two, the cause necessarily being geological and the effect biological. To test and discuss this concept the Paleontological Society held a symposium on "Distribution of Evolutionary Explosions in Geological Time," arranged by L.G. Henbest, then a micropaleontologist and stratigrapher with the United States Geological Survey. The following is the text of my contribution to that symposium as edited by Henbest and published in 1952. I have deleted two now unnecessary paragraphs but made no other significant changes.

The precise numbers of recognized orders, families, and genera in the graphs have of course been modified by later discoveries and research. We also now have somewhat better estimates of the lengths of the geological periods and epochs. For the most part, however, these do not seem to alter the interpretations drawn in any profound way. Probably the most important change in that respect is that many more Cretaceous mammals are now known. This eliminates the low point for Cretaceous mammals seen in Figs. 1–10 and 1–11 and the possible first peak for mammals in the Jurassic.

Since this paper was published much further attention has been given to the tabulation and interpretation of first and last appearances (sometimes but too loosely called "originations" and "extinctions") of named groups (taxa) of organisms in the known fossil record. These have often followed the concepts and used modifications of the methods here illustrated. When I wrote this paper emphasis tended to be on first appearances, but since then it has more often been on last appearances or on the relationship between the two. One interesting example among the many is cited below under *Another Relevant Reference.*

Text

The main purpose of the present symposium is to determine, first, whether there are periodic, general, essentially world-wide times of intensified evolutionary activity and, second, whether these correspond with similarly general and periodic increases in diastrophism, evolutionary events in the physical history of the earth's crust. The intention is to test a possible causal relationship, existence of which has commonly been taken for granted and is widely taught as fact to students of historical geology. The dogma has, however, been repeatedly questioned both on the biological and on the physical side. The choice of the topic for this symposium shows that doubts exist as to the validity of earlier conclusions on this subject.

This particular contribution is concerned with the biological aspect, only, of the problem and with the evidence of a single phylum of animals, the Vertebrata. The data for this phylum, although obviously incomplete, are exceptionally good both in the abundance of paleontological records and in the progress that has been made in their compilation and interpretation. The present aim is to summarize some of these data in such a way as to bring out any periodicity or other regularities that may exist in episodes of proliferation and of extinction among vertebrates. Discussion will then bear on the possible evolutionary significance of such regularities as are found and on their possible correlation with events in earth history.

The problem involves origins, increases, decreases, and extinctions of groups of vertebrates. The data are records of occurrence of the various taxonomic entities within each period (or, as may be practicable, epoch or age). Strictly speaking, the factual data do not include times of origin or of extinction, but only the times of first and last appearance in the record as known. These times of appearance certainly do not correspond precisely with real origins and extinctions, nor can the recorded increases and decreases be taken as accurately proportional to those that really occurred. In the present summary it is unnecessary to enumerate or evaluate the numerous sources of error, although these have been carefully considered and allowance has been made for them in deductions from the record. The data are taken as they are. First appearances in the known record are accepted as more nearly objective and basic than opinions as to the time when each group really originated.

The record thus compiled is clearly not random, as will appear below. It has a large measure of regularity and consistency that seems to attest correspondence with real events. In particular, well-marked trends of increase or decrease seem, with some exceptions and when taken broadly, to be significant even though they cannot furnish a precise measurement of the real rates. Sharply delimited highs and lows in recorded frequencies seem, with similar exceptions and precautions, to correspond closely with times of real maxima and minima in proliferation of the groups in question.

The numerical data consist of a census for various taxonomic categories known for various times in the geological record. Analysis of numbers of species was not attempted because of factors, relating mainly to sampling and to subjectivity, that make inferences from them excessively unreliable. Within the Vertebrata, records of orders and genera are most useful. Records of families have also been compiled and are included in some of the present graphs (Figs. 1–9, 1–10), but in general they add little to what is shown by the orders and genera. For each time and for each selected taxonomic level, figures in the tabulations (not here published as such) include total numbers, first appearances, last appearances, and continuations from preceding times. For the particular subject of this paper, first appearances are most significant because interest is focused mainly on times of maximum proliferation or episodes of so-called "explosive" evolution. It is also true that the total numbers, especially for genera, usually closely follow the general trends for numbers of first appearances, as can be seen in Fig. 1–8. Some broad data on last appearances (Fig. 1–12) and on totals and survivals (Fig. 1–13) are given but emphasis is here on first appearances.

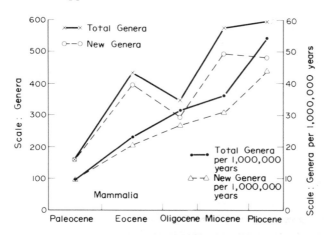

FIG. 1–8. Graphs of known numbers [as of 1952] of known genera and of first appearances of known ("new") genera of mammals in epochs of the Cenozoic. (The last, shortest epochs—Pleistocene and Recent—are not included.) Both first appearances and total numbers show nearly steady increase when scaled to approximate lengths of the epochs in years. This is affected by some sampling error, but is believed to be a real phenomenon.

Numbers of known taxonomic units are given by periods and not by shorter time units except for mammals, which are in some graphs (Figs. 1–8, 1–10) tabulated by epochs in the Tertiary. Some allowance must be made for the fact that the long time-units, although useful for purposes of broad summary, conceal some important details. Similar allowance is necessary for the conventionalization of placing each record at the middle of the corresponding time unit.

Even if the exceedingly short Quaternary period is omitted, the lengths of the geological periods or epochs differ by a factor of at least 3 to 1. (Few geologists will question that the Cambrian was three times as long as the Mississippian, for instance, or that the Eocene was about twice as long as the Oligocene.) With a constant rate of proliferation, the graphed values of first appearances would be much higher for a long period than for a short period. The fact that the lengths of the periods are not accurately established introduces another uncertain and subjective element, but comparability is better if allowance is made for variations in these lengths. In most of the graphs (Figs. 1–9 to 1–12), this is done by showing, not the total count for each period, but this count divided by the assumed length of the period in millions of years. This usually does, in fact, produce smoother and apparently more consistent graphs, as shown by the example of Fig. 1–8. The time scale at the bottom of each graph shows the relative assumed lengths of the periods.

The fossil record of the birds, which is still very poor, involves so many adventitious irregularities and would require so much inference and discussion that it is omitted here. Such as it is, this record (which has also been compiled and graphed but not published) is not inconsistent with conclusions drawn from other, better data for vertebrates as a whole. For all the groups, Pleistocene and Recent records are omitted, because the very different sampling conditions for those epochs would require lengthy discussion and adjustment of the data.

The curves of first appearances per million years for seven classes of vertebrates (Figs. 1–9, 1–10) show certain characteristics common to all or most of these classes. Each curve, whether for orders, families, or genera, rises to a relatively early peak, generally in the period of first appearance of the class or in the following period. Only for the family and generic curves of the Amphibia is the first peak as late as the second period after that of first appearance of the class. Mesozoic mammals are so poorly known that the reality of the low, recorded Jurassic peaks may be questioned, but better

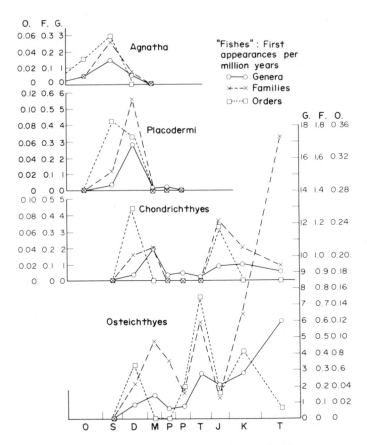

FIG. 1–9. Graphs of first known appearances per million years of orders, families, and genera of "fishes." The animals called "fishes" in the vernacular are usually placed technically in the four different classes here shown. O., F., and G. above the scales stand for orders, families, and genera. The time scale at the bottom indicates the relevant geological periods, oldest to the left, in sequence Ordovician, Silurian, Devonian, Mississippian, Pennsylvanian, Permian, Triassic, Jurassic, Cretaceous, Tertiary. (The Mississippian and Pennsylvanian are commonly united in a single period, the Carboniferous. The Tertiary comprises the Cenozoic epochs from Paleocene to Pliocene.)

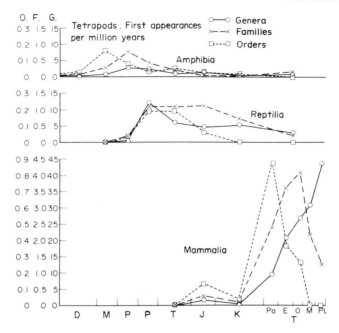

FIG. 1–10. Graphs of first known appearances per million years of the three classes of tetrapod vertebrates (those that were primitively four-footed). The method of construction and the abbreviations are as in Fig. 1–9 with the addition of the five epochs of the Tertiary period to the time scale for the Mammalia.

knowledge is more likely to emphasize these peaks than to eliminate them. Each curve tends to fall off rather rapidly from this first peak. In the period following the peak, there is a significant drop in almost every case. The only noteworthy exceptions to this are the ordinal and family curves for reptiles, which maintain the peak level for two and three geologic periods, respectively. Some but as yet insufficient, further analysis strongly suggests that this apparently exceptional feature in the reptilian record is an artifact caused by the coarseness of the time scale for the present graphs. If scaled to epochs or ages, it is highly probable that there was a drop in rate after the Permian peaks and that the apparent continuation of these peaks in the Triassic represents, in fact, a new rise in each curve. The reality of the Cretaceous drop in the mammalian curves is open to considerable question, yet this does agree with the usual trend of the other curves.

After the first peaks, the curves for the two earliest and most primitive classes decline more or less steadily to extinction (Placodermi) or near it (Agnatha, with a few living relicts but no fossil record after the Devonian). The Chondrichthyes have a clear-cut second peak on all three curves, and the Osteichthyes have not only a second but also a third peak. The Amphibia have slight but probably real second peaks for orders and for genera. There is no second peak in the present graphs for the Reptilia, but (as noted above) this is probably an artifact and second peaks probably did occur at each taxonomic level during the Mesozoic. If the Jurassic peak for mammals is real, then their very sharp and unquestionably real Tertiary peaks are to be considered second peaks.

Omitting the dubious first (?) mammalian and second (?) reptilian peaks, the distribution by periods is as follows:

TABLE 1–9

	Ordinal peaks	Family peaks	Generic peaks
		Osteichthyes 3	Osteichthyes 3
Tertiary	Mammalia (2?)	Mammalia (2?)	Mammalia (2?)
Cretaceous	Osteichthyes 3		Chondrichthyes 2
Jurassic	Chondrichthyes 2	Chondrichthyes 2	(Amphibia 2?)
Triassic	Osteichthyes 2	Osteichthyes 2	Osteichthyes 2
	(Amphibia 2?)		
Permian	Reptilia 1	Reptilia 1	Reptilia 1
Pennsylvanian		Amphibia 1	Amphibia 1
Mississippian	Amphibia 1	Chondrichthyes 1	Chondrichthyes 1
		Osteichthyes 1	Osteichthyes 1
Devonian	Chondrichthyes 1		
	Osteichthyes 1	Placodermi	Placodermi
Silurian	Agnatha		
	Placodermi	Agnatha	Agnatha

The spaces are almost completely filled. In *every* period after that in which the phylum first appears (Ordovician), some class of vertebrates was in a phase of markedly rapid proliferation of new groups. Moreover this is true for each of the three taxonomic levels graphed except for orders in the Pennsylvanian and families in the Cretaceous. The apparent exceptions have little or no significance. A *class* of vertebrates (Reptilia) first appeared in the Pennsylvanian, and rates of appearance of new families were high in the Cretaceous (for Chondrichthyes, Osteichthyes, and Reptilia) even though still higher rates in adjacent periods keep these from appearing as peaks on the broadly generalized curves of the present graphs.

Thus, in the words of the original proposal for this symposium by L.G. Henbest, *within* a single class (and the same is true for smaller subdivisions not here graphed) there is definitely a tendency for "proliferations . . . to be segregated in particular epochs." As between different groups or for the Vertebrata as a whole, however, there is certainly no tendency for their times of most rapid diversification to coincide. The various segregated times of proliferation for each class or lesser group are so scattered through geologic time that one or more occurs in every period pertinent to this history. In this sense, "the proliferations are not segregated but are distributed more or less evenly through time."

This distribution of peaks of differentiation of new groups through time is more or less even, but it is not random. There are evident systematic relationships between the various peaks. Possible interpretations of this fact require further discussion, which follows on later pages.

Within each class, there is a strong tendency for rates of appearance of new genera to rise and fall in the same sequence as the rates for orders, but to do so at later times. In the ten most clear-cut peaks seen in the graphs (Figs. 1–9, 1–10) a generic peak occurs one period later than the ordinal peak in six cases. The four apparent exceptions, which have ordinal and generic peaks in the same period, are probably not really such in any case but are artifacts of the presentation on a coarse time scale. In each case the generic peak was probably later than the ordinal peak by a considerable time and yet occurred in the same period. This is positively established for the Mammalia and is very clearly shown in the graph of epochs (Fig. 1–10, bottom) rather than by periods. Ordinal and generic peaks occur in the same period, the Tertiary, but the former in the Paleocene and the latter in the Pliocene, with an interval of at least 45 million years between them.

The evidence is that a generic peak regularly (usually and perhaps always) occurs some 25 to 50 million years after an ordinal peak. In the case of the mammals, graphed with a more refined time scale, the family peak falls between those for orders and for genera, as might be expected. On the period scale, the family curve tends to parallel the generic rather closely and simultaneously. This perhaps reflects the classifier's frequent habit of treating a family simply as a bundle of genera, so that more genera automatically mean more families, whereas the number of orders recognized is largely independent of the number of genera. However that may be, further consideration of families can add little to conclusions from the relationships of appearance rates in orders or genera.

As vertebrate classification has been worked out and is reflected in these data, large categories such as orders represent basic and new adaptive types. Lines of descent from these normally diverge and involve adaptive radiation, but within the scope or building on the basis of the ancestral ordinal type. Families, genera, and other lower taxonomic units categorize the lesser adaptive types. They also include the many more or less adventitiously isolated groups, which reduplicate a type in a variety of localities and environments and which fill in many ecological niches within broadly similar life zones. The vertebrate record shows that the peak of splitting into basic types, as exemplified in rate of appearance of new orders, considerably precedes the similar peak of splitting into lesser groups, such as genera. When genera are appearing at their highest rate, the rate of appearance of orders is dropping rapidly. As regards total numbers of taxonomic units, also, in a given group (as a rule) when the number of orders is highest, the number of genera is lower than it later becomes. Past the ordinal peak, the number of orders tends to decrease as the number of genera increases.

Now, in studying the supposed phenomenon of "explosive" evolution and testing possible correlation of this with physical events, when does the "explosion" occur? By definition, the "explosion" is a relatively abrupt increase in numbers of taxonomic units, or in their rate of appearance. In absolute numbers, genera and species are most numerous and they are also at the level of usual taxonomic study. The "explosion" is commonly taken to be an increase at this level. But peaks of incidence of new genera follow those of new orders with such regularity in the record here presented that it seems inescapable that the fundamental event, in some sense the cause of the expansion, is the earlier ordinal differentiation. Then seeming correlation of generic "explosion" with contemporanious physical events would appear to be purely adventitious, because the generic expansion was the outcome of events of some 25 or 50 million years earlier.

In fact, it becomes impossible to pin down the essential point of this chain to any particular expansive phase in it. A most fundamental adaptive type categorized as a class appears and in 25 to 50 million years its rate of ordinal differentiation reaches an "explosive climax." In another 25 or 50 million years the rate of generic differentiation reaches a similar but quite distinct climax. The whole process seems to be a biological, evolutionary sequence affecting all vertebrates throughout the history of the group. It naturally occurs against the background and within the limitations of the physical environment, but it seems impossible even to make a start at realistic correlation of so regular and continuous a process with intermittent tectonic episodes. The occurrences of second and third climaxes within a given class fall into exactly the same sort of sequence, for without exception they also are initiated by emergence of new adaptive or structural types which simply do not happen, in our classifications, to be formalized at the rank of classes. Once they arise, a sequence of ordinal and generic highs follows in due course as it did after the class as a whole arose.

The second peaks of the Chondrichthyes involve the appearance of "higher" or "modernized" sharks and rays: galeoids, squaloids, and batoids. Those of the Osteichthyes followed the appearance of the subholostean–holostean structural lines. The third peaks of the Osteichthyes follow the appearance of teleosts, now the dominant fishes.

In the probable but not absolutely established occurrence of two peaks of reptilian differentiation, one late Paleozoic and one early Mesozoic, the first involves the cotylosaurian–synapsid level (the two groups being adaptively and structurally very similar although their different fates have caused their radical separation in the usual classifications). The second set of peaks follows in regular sequence after appearance of the archosaurian level of organization in the Triassic.

A classic example of supposed "explosive" evolution and its correlation with diastrophism is the "explosion" of mammals supposed to initiate the Cenozoic and to be simultaneous with if not caused by the Laramide Revolution. But in fact the most basic event, the origin of placentals, occurred sometime well before the end of the Cretaceous. Most of the orders of early Cenozoic mammals did not appear at the Cretaceous–Paleocene boundary, but straggled in over a span of some 20 million years. The rate of appearance of new genera was low in the Paleocene and it did not reach its climax, its most truly "explosive" phase, until the Pliocene, perhaps 60 million years after the end of the Cretaceous. When did the "explosion" occur? Certainly not at the Cretaceous–Paleocene boundary, and claimed relationship to the Laramide Revolution must surely be viewed with suspicion.

The most fundamental events of all within the span of vertebrate history, the origins of the classes, are not "explosions" in any sense, since each class is rare and little varied when it first appears, but these most major first appearances also follow a biologic sequence which seems to have no possible correlation with tectonic episodes dividing the periods and eras. Such appearances do not regularly occur at the beginning of periods. The Agnatha appear in the middle (or possibly late) Ordovician [they are now known from the early Ordovician and probably late Cambrian—Note added in 1979], Placodermi in late Silurian, Chondrichthyes and Osteichthyes in early Devonian, Reptilia in middle or late Pennsylvanian, Aves and Mammalia in middle Jurassic. [Mammals as currently defined are now generally recognized as known from the late Triassic—Note added in 1979.] Although some, at least, of these classes may have originated considerably before their known appearances, it would be purely arbitrary to postulate origins at times of preceding diastrophic climaxes. Moreover no class of vertebrates and few of the major structural grades within the classes appear during or immediately after the two greatest tectonic episodes of the span of vertebrate history, the Appalachian Revolution and the Laramide Revolution.

The climaxes or peaks of rate appearance of new groups among the vertebrate classes form a fairly regular progression which clearly is not random. Or rather, they form two progressions, one for the primarily finned, aquatic types, the "fishes" (Classes Agnatha, Placodermi, Chondrichthyes, and Osteichthyes), and one for the primarily footed, amphibious to terrestrial types, the tetrapods (classes Amphibia, Reptillia, and Mammalia, with the Aves, not here discussed, forming a separate major ecological type of their own.)

The succession involved here seems, again, to be evolutionary in a sense primarily biological and I see no clear evidence for correlation of its main events with purely physical changes such as mountain-building. The most probable interpretation seems to be that at intervals some branch from an earlier radiation reaches a superior adaptive status which enables it largely (seldom completely) to replace earlier forms living over somewhat the same environmental range. In taxonomic terms, a new class, subclass, or superorder arises and the sequence of ordinal down to generic and specific "explosive" phases follows, as outlined above.

The sequence of such phases in the Agnatha, Placodermi, and Chondrichthyes plus Osteichthyes is quite clear from Ordovician to about the end of the Paleozoic. (See Figs. 1–9, 1–11.) The ordinal climax of each group coincides approximately with the generic climax of the preceding group. In the Pennsylvanian and Permian, with no radically different or successional new adaptive type on the scene, the rates of appearances of new groups in Chondrichthyes and Osteichthyes fall to low points and evolution of "fishes" becomes relatively stagnant. In this part of their history, Devonian through Permian, Chondrichthyes and Osteichthyes are not successional to each other but simultaneous and closely parallel in the rise and fall of rates of appearances. The two, together, and not either separately or each in succession, seem to replace the earlier

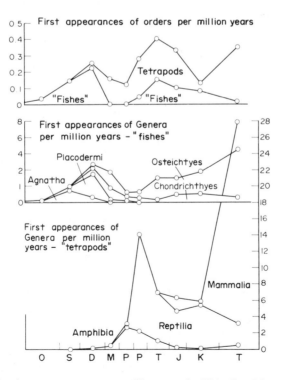

FIG. 1–11. Graphs of first known appearances per million years for "fishes" and for tetrapods. The time scale and conventions of construction are as in Fig. 1–9.

"fishes". There is, in fact, evidence that most Chondrichthyes were then marine and most Osteichthyes freshwater, so that they parceled out the aquatic domain between them.

In the Mesozoic and Cenozoic, as many Osteichthyes became marine, there is no longer simultaneity in "explosions" of these two classes, but a complex succession see-sawing from Osteichthyes in the Triassic to Chondrichthyes in the Jurassic to Cretaceous and back to Osteichthyes in the Cretaceous to Recent.

Succession of "explosive" episodes among the tetrapods is simpler and rather obvious throughout. (See Figs. 1–10, 1–11.) The only complications in the Amphib-ian-Reptile-Mammal sequence are in relatively minor points and need not be discussed in this broader summary.

Although this study is mainly devoted to episodes of appearance of new groups of vertebrates, some of the more striking, related features of total expansion and contraction may be briefly mentioned.

Changes in total numbers of taxonomic units between any two times depend on a balance of four factors:

(1)   splitting up of single groups into two or more;
(2)   transformation of single groups to the point recognized as origin of new groups at the same taxonomic level;
(3)   survival of groups without change sufficient for recognition at the given taxonomic level;
(4)   disappearance of groups, without known descendants.

As the data are usually set up, these factors are not clearly distinguishable and cannot be accurately analyzed. Such analysis can be made (although even here with some difficulty in categorization) for a few, small, exceptionally well known and well studied groups, such as the horses. For most groups and particularly for large groups such as those involved in the present summary, the data do not as yet exist, or are so incomplete that they would not repay the great labor of accurate analysis. In broader studies at present, therefore, the data used reflect the interplay of these factors but do not ade-quately indicate the contribution of each. "First appearances" include contributions from both (1) and (2), without distinction, although (1) increases the total and (2) does not. In a continuous sequence, the number of first appearances due to (2) increases with the length of the time unit used, although this misleading effect can be eliminated if first appearances are plotted relative to time, as in most of the graphs here presented.

"Last appearances" include contributions from both (2) and (4), again without dis-tinction, although those due to (4) decrease the (subsequent) total and those due to (2) do not. Again, the last appearances due to transformation, (2), tend to be proportional to the time covered and their real significance is not evident unless they are plotted relative to time. First and last appearances due to (2) are equal in number and balance each other exactly. The last appearances in any given time unit affect the total not in the same but in the following time unit.

Survivals from one time unit to the next, (3), tend to maintain the total level without change and are readily tabulated but are of obscure and variable significance. The number of survivals obviously decreases greatly as the length of the (essentially arbitrary) used time unit increases, and this confusing effect cannot be eliminated simply by plotting survivals relative to time. (In fact, this merely exaggerates the defect; plotting the reciprocal figure does give a more reliable result, but one still difficult to comprehend and interpret.) The tabulation or plotting of survivals has little or no clear significance unless the time unit is considerably shorter than the average time span of the taxonomic unit used. Since the average span of vertebrate genera is less than that of a geological period, study of survival in genera is not useful on the scale of broad summary used in the present short contribution. Orders of vertebrates do tend to persist considerably longer than one geological period and so study of their survival from one period to another does have some usefulness (see Fig. 1–12 and text, below).

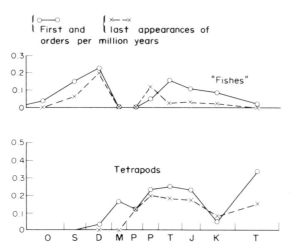

FIG. 1–12. Graphs of first and last known appearances of orders per million years in "fishes" and in tetrapods, constructed as in Fig. 1–9.

Figures for the last appearances of genera or other small taxonomic units tend, unless the time scale used is more refined than is generally practicable for large groups, simply to run parallel to the figures for first appearances. This has two causes. First, the last appearances due to (2) in the list of factors given above are always equal in number to first appearances from the same cause and occur, theoretically, at the same point in time. Second, last appearances representing true extinction, (4), tend to be proportional to total numbers and to be accelerated by increase in first appearances. There is normally a time lag in this case, but in the case of smaller taxonomic units it may be too short to be clear unless the data are quite exceptionally good and a brief time unit can be used.

Even at the level of orders, figures for first and for last appearances tend to run parallel (Fig. 1–11), although some significant interplay can be seen. For "fishes" the Devonian was the time of maximum rates both for first and for last appearances. This is a reflection of replacement. The peak of first appearances is due to the rise of the Chondrichthyes and Osteichthyes and the peak of last appearances is due to waning of the Agnatha and Placodermi. In the Mississippian and Pennsylvanian both first and last appearances of orders of "fishes" dropped to a minimum. (For genera, this minimum was Pennsylvanian to Permian, see Fig. 1–9.) During this time, the successful forms in the Devonian competition simply tended to survive with relatively little change. In the Permian, last appearances increased markedly without equal increase in first appearances; this is the only period in which last appearances were more numerous than first appearances of orders of "fishes." In combination with the drop of last appearances and rise of first appearances in the Triassic, this reflects a radical turnover of basic types of "fishes" from Permian to Triassic. After the Triassic there are no marked changes. Both first and last appearances of orders dropped steadily (and first appearances and totals of genera rose steadily) as the broad structural grades introduced in the Triassic survived and proliferated at lower taxonomic levels.

Among tetrapods, the equal first and last appearance rates of the Pennsylvanian and high for both in the Permian and Triassic accompany rapid turnover among early amphibians and reptiles and the beginning of replacement of some of the former by the latter. The drop in both rates in the Cretaceous, with last appearances becoming for the only time more frequent than first appearances, followed by a rise of first appearances far above last, repeats in geologically briefer time and with some differences the sort of episode seen among fishes in the late Paleozoic and early Mesozoic.

The total of known groups at any time is of course the sum of first appearances and survivors from the preceding time unit (Fig. 1–13). The number of survivors, in turn, is the total for the last period minus last appearances in that period. As regards orders, the course and balance of these various factors and their results are sufficiently shown in Figs. 1–11, 1–12.

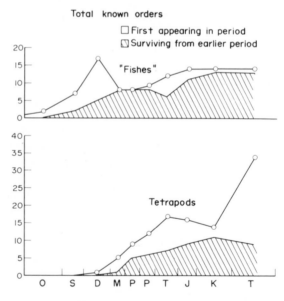

FIG. 1–13. Graphs of total absolute numbers of known orders of "fishes" and tetrapods. The crosshatched areas represent orders surviving from earlier periods. The time scale is as in Fig. 1–9.

In view of the picture as a whole, several periods of major evolutionary activity of different sorts stand out. The most striking of these seem to be:

Devonian: Rapid turnover from earlier to later major groups of "fishes." High extinction rates for older types. Extremely high basic origination rates for many sorts of "fishes," to be followed in subsequent periods by their diversification in lower taxonomic categories. Origin of tetrapods.

Permian to Triassic: Among both "fishes" and tetrapods, strongly pronounced turnover, with high last appearance rates in the Permian, at least, and high first appearance rates in the Triassic. (These major changes concern mainly the Osteichthyes and the Reptilia.)

Cretaceous to Tertiary: No definite climax or crisis among "fishes," but strongly marked tetrapod turnover invovled in decline of the reptiles and rise of the mammals.

It has long been realized that these were major episodes in vertebrate history and the present conclusion only confirms and to some extent quantifies what was already well-known. It should, however, be emphasized that these episodes are of long continuation and that they are not sharply defined in time, even as to their beginnings. It has, for instance, already been stressed that the essential features of the Cretaceous–Tertiary crisis cannot really be localized just at the boundary between those periods. The episodes are parts of a long and essentially continuous process.

The Permian–Triassic and Cretaceous–Tertiary major evolutionary changes do coincide with era boundaries, or, rather, the accepted era boundaries occur within the long and vaguely defined span embracing these evolutionary events. The era boundaries, in turn, coincide with mooted major tectonic episodes, the reality or true nature and extent of which are for others to judge. There may be some degree of causal connection within the complex totality of evolutionary processes involved and it is unlikely that physical events played no part in these evolutionary episodes. The conclusion that theirs was the decisive and necessary part is not supported by the biological evidence and may be gratuitous. It is also to be remembered that the era boundaries were established in large part on paleontological grounds, so that their coincidence with evolutionary events is not really, in another sense of the word, a coincidence.

In certain particular cases, strong or even startling evolutionary effects can be clearly traced to physical, geographic and tectonic causes. Completion of the Central American land bridge in the Pliocene was followed by geologically sudden and radical changes in the fauna of South America. Rates of extinction rose rapidly among native South American groups of mammals, particularly, and rates of diversification or proliferation became extremely rapid among groups introduced from North America. Similar results followed the rise of the Bering bridge between the Old World and the New in the late Eocene after long isolation through the middle parts of the period. In such cases the correlation of biological and physical events is obviously real. It is clear that such events have repeatedly occurred and constitute a large and important part of the history of life. They do not, however, as far as can be seen, account for the origins of important new types of structural grades of animals. The episodes of proliferation that follow them are also of secondary importance in the whole picture of earth history. Even when continent-wide, they are local with respect to the earth as a whole, and the proliferation is on low taxonomic levels, dispersion into new living spaces generally by replacement of older, ecologically similar, and often more or less closely related previous residents.

The physical concomitants of such episodes are also and still more strictly local. Relatively slight local warping of the crust in a small area of Central America or of the Bering region, for instance, is all that need occur. No widespread or major diastrophic crisis is required or is evidenced. On the other hand, world-wide and revolutionary orogeny and diastrophism could occur without having any particular results of this sort, because it need not affect the small areas that happen to be crucial. It is significant that the two examples given above (and many others that could be added) did not occur at times of disturbances or revolutions usually recognized by adherents of the theory of periodic world-wide diastrophism and did not occur at or shortly after conventional epoch or period boundaries.

It is probable that such events are sometimes involved in episodes that are important details in the broader picture of biological history. This may be true, for instance, of the changes used in fixing the Cretaceous–Tertiary boundary, especially disappearance of the last dinosaurs and appearance of a meager vanguard of Cenozoic mammalian types in the (few and local) collecting regions where these facts can be ascertained. The geographic event may then have determined a useful time datum and may enable us to draw a boundary in a given region. Still that is not evidence that the geographic event itself was more than a local crustal warp someplace or other, or that it did more than set the precise time of evolutionary events that would in any case have occurred at about that time in the sweep of the long, continuous evolutionary process.

The evidence summarized in this paper is consistent with the view that most of the broad features of vertebrate history might have been much the same if the earth's crust had been static (providing that the surface remained sufficiently varied and with large connected and nearly connected land and sea areas). Crustal movements may have had essential roles only as regards details of timing and of distribution, important details in some cases but still only details. The older and still perhaps more common belief in causal synchronism of periodic world-wide evolutionary and diastrophic episodes is certainly not disproven, but the evidence runs rather more against than for it. The most likely points at which physical events may have had decisive influence are the extinctions of aquatic vertebrates around the end of the Permian and of terrestrial vertebrates around the end of the Cretaceous.

There is a well-marked periodicity in vertebrate history, especially as regards successive peaks and valleys of high and low proliferation of new groups, in one class after another or sometimes within one class at different times. Taken as a whole, these peaks are rather evenly distributed in geologic time and have no clear segregation that could correspond with periodic crustal crises. Within any one major group, the peaks at different levels of adaptive differentiation spread very widely in geologic time, over spans up to 100 million years or more, which makes difficult or impossible their connection with particular tectonic events more sharply defined, as at period boundaries. The regularity of this long process is, moreover, suggestive of continuity of evolutionary forces acting regardless of periodic physical events despite the periodicity of some of the results. The "explosions" in evolution are not really isolated occurrences or brief even in the sense of a few million years of geologic time. They seem, rather, to be merely regular, shorter episodes or manifestations in a more basic process which is really secular and not periodic.

The origins of classes, orders, or other taxonomic units, either singly or in the accelerated clusters of "explosions," do not, in the record, show any apparently causal correlation with epoch or period boundaries, beyond the extent to which these boundaries have themselves been set to correspond with some convenient evolutionary marker. The important Paleozoic–Mesozoic boundary, for instance, occurs within and near the climax of a radical faunal change, which is used to define that boundary, but it does not coincide with, or even approximate, the time of origin of any basic new type of vertebrate or any peak in proliferation of some group. It is marked mainly by negative, not positive, evolution, by extinction. The subsequent positive faunal changes may have been stimulated by preceding extinctions but may also have been the outcome of events that occurred much earlier or indeed of a whole chain of events back to the origin of the vertebrates or of life.

As far as this evidence goes, then, diastrophism and other physical events continue to hold a place in the totality of factors that are, inseparably, the cause of evolution. They cannot be promoted to a role as leading and distinct cause of the major features of vertebrate evolution. The related but different and more particular theory of accelerated evolution during rhythmic or periodic diastrophic disturbances or revolutions deserves, at best, a verdict of "not proven." For the vertebrates, at least, it may be proper to add "improbable."

Source

(1952) Periodicity in Vertebrate Evolution, *Jour. Paleontology*, Vol. 26, pp. 359–370.

Another Relevant Reference

Newell, N.D. (1967) Revolutions in the History of Life, *Geol. Soc. Amer.*, *Special Paper* 89, pp. 63-91.

2

Morphology, Homology, and Function

AMONG the most basic aspects of study of the fossil record are the examination, description, and illustration of the morphology of fossils. That sometimes involves little more by way of method than naked eye scrutiny, identification of anatomical features, their description in the appropriate vocabulary, and illustration by any graphic (including photographic) method. However, much more varied and specialized methods are often, indeed now usually, involved. A next step is to compare different fossils with a view toward determining their affinities and classifying them. Another aspect of anatomical comparison is the determination of equivalences among varying structures, which involves the concept and recognition of homologies and analogies. Finally, as far as this chapter is concerned, there are problems and methods of studying the functions of anatomical parts.

In this chapter just a few examples involving methods applied to each of those kinds of studies are given. Most of the methods here noted were unusual in such applications at the time of the publications from which extracts are here given. A few were then new, as far as I know. Procedures logically following such anatomical observations and interpretations involve subjects treated in the following chapters.

A Problem of Black on Black

Comment

In 1857 Ebenezer Emmons, a professor at Williams College and Director of the North Carolina Geological Survey, published a cursory description of two tiny jaws that had been found in a coal mine in North Carolina. These at once became famous as the oldest known mammals. Emmons called them both *Dromatherium*, which means "running beast," although it is not quite clear why Emmons thought the name appropriate. In 1887 Henry Fairfield Osborn made a somewhat more careful study of the specimens, which he thought to represent two distinct genera, and he named the second one *Microconodon*, "tooth (with) small cones," correctly descriptive but not very distinctive. No further specimens of those genera have yet been found. Well into the 1920s they were still accepted as among the earliest, if not the very earliest, of known mammals. In 1925–26 I was writing a doctoral dissertation on Mesozoic mammals, those already present during the long span that is called the Age of Reptiles. I therefore undertook to restudy these supposedly Mesozoic mammals, one specimen (*Dromatherium*) then (and now) at Williams College and the other (*Microconodon*) in the Philadelphia Academy of Sciences. I found that the previous descriptions and figures had been quite inaccurate and that these animals were almost certainly mammal-like reptiles and not mammals in a stricter sense.

That conclusion has been generally accepted ever since, and it is not the reason for here including an extract from the paper that I wrote as a graduate student. The most serious problem that I encountered was the fact that the specimen of *Dromatherium* was just the same color as the dead black of the coal in which it was embedded so that its detailed structure could hardly be, and hitherto had not been, properly made out. The solution to the problem was provided by the fact, as related in the following extract, that the specimen fluoresced in ultraviolet light and the coal did not. As far as known, that was the first use of fluorescence in paleontology. The method was soon thereafter independently discovered by two German paleontologists, Miethe in 1927 and by Wagner in 1928. Since then it has been widely used, and it is included in handbooks of paleontological techniques.

There are of course many other aids to observation now available for study both of fossil and of recent organisms. At present the most advanced and most interesting is the scanning electron microscope, which is not further mentioned in this book because I have never had occasion to use it, although some of my students have. I did use in my dissertation a method of illustration that was then rare in paleontology, or possibly new in this application: stereophotography. It is now very widely, almost routinely, used in illustrating fossil vertebrates and some other fossils. An apparently unique application that I tried out was a long strip of photographs in which any two adjacent ones were stereoscopic and the whole series went all around the specimen so that moving a stereoscope along the strip had the effect of turning the specimen through 360°. The Hungarian paleontologist Kálmán Lambrecht published three stereoscopic pairs from my strip in a book on fossil birds in 1933, but as far as I know the 360° stereograph idea has not been mentioned in print before now and no one has made another.

Text

Since the new descriptions and figures here given differ in some important respects from the very careful and universally accepted ones of Osborn, it seems necessary to give an account of the methods employed in order that the reader may be in a position to judge as to their reliability. Much better optical instruments are made today than were available forty years ago and very fine equipment was placed at the writer's disposal. In both Williamstown and Philadelphia excellent high power binocular microscopes were used. *Microconodon* proved to be quite clearly shown, despite the fact that it is smaller and more broken, and examination under various powers of magnification with the light falling from different angles was found to leave no question as to its characters. Direct and diffused sunlight gave entirely adequate illumination. *Dromatherium* offered greater difficulty as it is much more obscure. The bone and teeth are the same color as the black coal in which they are imbedded and the boundaries are very difficult to make out in many places. This black color and a certain vagueness of form at several points also adds difficulty to interpreting the surface of the specimen itself. Direct and diffused sunlight and incandescent electric light were used with all powers of the microscope.

Through the cooperation of Professors Brinsmade and McElfresh of the Department of Physics at Williams College, it was also possible to use a small and very convenient laboratory mercury arc in quartz. Various ray filters were employed in the endeavor to bring out some possible subtle difference in pigmentation of the specimen and matrix, but they seemed to be equally black with all colors of light. It was found, however, that the simple use of the unmodified light from the arc produced a remarkable result. The ultraviolet rays in which this radiation is unusually rich set up a bluish fluorescence in the enamel of the teeth which, while not strong, was sufficient to distinguish their boundaries with the matrix very clearly. In this way it was possible, for example, to show that the anterior teeth were all drawn much too slender by Osborn, only the central part of each tooth being recognized as such. Such points as still remain doubtful are so stated in the descriptions which follow.

In making the drawings a camera lucida was not used, as the boundaries are so obscure under the best of conditions that the use of a camera would seem more likely to introduce minor inaccuracies than to eradicate them. Every part of the jaws and dentitions was measured with the greatest care and the dimensions transferred directly to the drawing by the use of proportional dividers, a procedure involving absolute accuracy of the order of less than 0.1 mm and relative accuracy (that is in comparing the different parts of the same jaw) probably somewhat greater even than that. Original drawings were made by the writer in pencil and these have been exactly copied in ink by Mrs Louise Nash for publication in the present paper. (See Fig. 2-1.)

FIG. 2-1. The right lower jaw (dentary bone) of the supposed mammal, probable mammal-like reptile, *Dromatherium*, with teeth somewhat damaged. The drawing was made as described in the text. The total length of the specimen is 23.2 millimeters.

Source

(1926) Mesozoic Mammalia. V. *Dromatherium* and *Microconodon*, Amer. Jour. Sci., 5th series, Vol. 12, pp. 87-108.

Other References

KUMMEL, B., and D. RAUP, editors (1965) *Handbook of Paleontological Techniques*, W.H. Freeman, San

Francisco and London. (Fluorescence under ultraviolet, pp. 351–360; stereophotography, pp. 428–429; the dust jacket and binding of this work also illustrate fluorescence in a fossil.)

LAMBRECHT, K. (1933) *Handbuch der Palaeornithologie*, Gebrüder Borntraeger, Berlin. (Bits from the 360° stereograph, Fig. 41 on p. 97.)

"Braincasts"

Comment

The anatomy and evolution of brains is of special interest to scientists who, as humans, consider their own species to represent a pinnacle of such evolution. The brain itself, soft tissue that quickly decomposes after death, is not preserved in fossils. However in birds and mammals the bony cranium, which can be preserved, fits so closely to the brain that the brain cavity or endocranium gives a fair representation of the size and shape of the brain although often lacking fine details. Thus what are loosely called "braincasts" of fossils are now more often called "cranial endocasts" by paleontologists. Even this is not strictly accurate because in fact what are studied are molds, not casts, of the cavity. They are studied because they approximate what an actual cast of the long-lost brain would look like if we had one. In addition to that approximation, a good endocast (in the now usual terminology) also gives useful data on the cranial nerves and some other details. In fishes, amphibians, and reptiles the brain does not fill the cranial cavity, but an endocast does give some, here less detailed, knowledge of the brain and also of the exits of cranial nerves and of various other openings (foramina).

When skulls become buried in sediments their cavities often are filled by clay, silt, or sand which later becomes hardened and forms a natural internal mold or endocast, which may be exposed and studied. More often now the cavity is cleaned out and a mold of it taken by a layer of latex, which is flexible enough to remove even through a relatively small opening and then resumes its shape to be cast in some harder and more durable material.

The study of cranial endocasts, mostly mammalian, was already extensive in the 19th century and is still active and widespread today. As an example a figure and comparative description of an endocast of the South American early Eocene genus *Notostylops* is here extracted from a longer paper. In order to illustrate the technical detail and the comparative method, the treatment of this one specimen is quoted almost in full. Comparison is with an endocast of the North American Eocene condylarth *Phenacodus*. The condylarths were the most primitive hoofed mammals and their brains probably were more or less near an ancestral condition for the ungulates in general. *Notostylops* was an early member of an order, Notoungulata, that became the dominant group of herbivores in South America but finally became totally extinct. The comparison of *Notostylops* with *Phenacodus* was designed to show whether and how the early, primitive notoungulate brain had evolved beyond the ancestral condylarth stage. The results show that the notoungulate brain was somewhat distinctive but still quite primitive in the early Eocene.

Two other examples different in kind are here shown in figures without quoting the texts that originally accompanied them. *Ptilodus* is a North American Paleocene mammal of the extinct order Multituberculata. The reconstruction here shown in Fig. 2–2 is based on a crushed and broken skull collected in a fossil quarry in Montana. The various

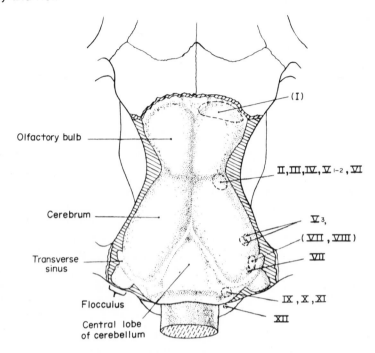

FIG. 2–2. Reconstruction of part of the skull of the Paleocene multituberculate *Ptilodus montanus* in dorsal view with the cranial roof represented as cut away to show an endocranial cast. The Roman numerals designate the cranial nerves and they are keyed to outlines of the canals or foramina by which they leave the cranium. The numerals in parentheses are tied to the internal and the others to the external apertures. The cut cylinder at the bottom of the figure represents the beginning of the spinal cord.

fragments of the cranium were freed from the surrounding rock and then although it was not practicable to make an actual endocast a drawing of one, along with some other features of cranial anatomy, was put together from study of both inner and outer surfaces of the bits of bone. The reconstruction indicates a brain radically unlike that in any other group of mammals. Other specimens have since confirmed the presence of these unique features in the multituberculates.

Fig. 2–3 illustrates the anatomy of one part of a natural endocast of a small mammal-like reptile from the Triassic of South Africa. In this specimen the sediment filling the skull cavities became so firmly hardened that when the surrounding bone was dissolved or weathered away, the filling was preserved and exposed in full relief.

Text

I have elsewhere described a partial natural brain-cast of *Notostylops*. The natural cast (Museo Nacional, Buenos Aires, No. 10506) had the cerebellum very poorly preserved and did not reveal the ventral surface or any nerve exits, and the present specimen is therefore much more satisfactory, although also imperfect in some details. The two specimens differ somewhat, most obviously in the greater postcerebral length of the present cast.

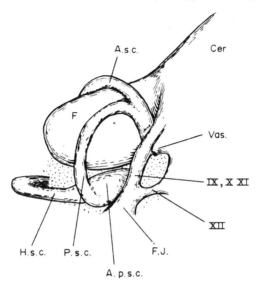

FIG. 2–3. Sketch of a natural internal mold of some of the cavities and canals in the skull of a mammal-like reptile, *Nythosaurus*. This includes much of the left ear region seen from behind. This includes the three semicircular canals, organs of equilibrium, and the cavity for the bulb (ampulla) of the posterior semicircular canal (A.p.s.c.). The Roman numerals designate canals for cranial nerves and Vas. is the stub of a canal for a blood vessel. Cer. is part of the top of the cavity for the central part of the hindbrain (cerebellum) and F. is the cavity for a lateral projection (flocculus) from the hindbrain.

The specimen here described and compared is an artificial endocranial cast of a skull of *Notostylops*, Amer. Mus. No. 28614, found by me in Cañadón Vaca, Chubut, Argentina. It is somewhat crushed dorso-ventrally, most of the left side is lacking, and it was not possible to free the olfactory bulbs, which were very difficult to reach with tools and were filled with barite crystals. Yet all the essential characters posterior to the olfactory bulbs are well shown with partial restoration in Fig 2–4.

The estimated total length of the skull is 115 mm, and of the brain, exclusive of medulla, 55 mm, giving an index of 48. Even apart from the influence of brain structure, this larger index does not necessarily indicate a larger effective brain capacity than in *Phenacodus*. *Phenacodus primaevus* is twice as large in linear dimensions, and larger animals have relatively smaller brains, other things being equal, and it also has an elongate snout, tending to give a low index while that of *Notostylops* is abbreviated, tending to give a high index. The effective brain size, if a name may be applied to a factor so elusive and so impossible to separate from other and even more important factors of brain development, was probably about the same in the two animals.

The serial arrangement is about as in *Phenacodus* and the flexure is, if anything, slightly less. Olfactory bulbs and cerebellum are as fully exposed, and the midbrain apparently even more so. The ratio of olfactory bulbs, cerebrum, and cerebellum was about 2:4:2, the difference from the 2:4:3 of *Phenacodus* being due to the large dorsal gap between cerebrum and cerebellum, part of which doubtless belongs to the cerebellum although not included in its measurement.

The rhinencephalon is very strongly developed, as in *Phenacodus*. In the Buenos

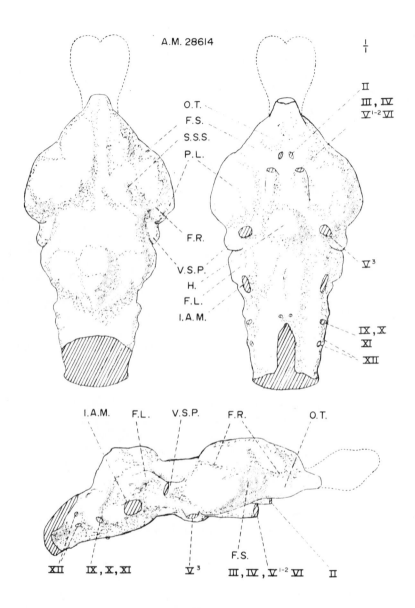

Fɪɢ. 2–4. Artificial endocast of the cranium of *Notostylops*, a primitive Eocene notoungulate mammal of South America in dorsal (upper left), ventral (upper right) and right lateral (lower) views. The Roman numerals represent the inner ends of canals for cranial nerves. The technical names for the structures labeled by initials are: Fl., flocculus (a lateral projection of the hindbrain); F.R., rhinal fissure (indicating an inward fold of the brain); F.S., sylvian fossa (another brain feature); H., filling of the hypophyseal fossa (where the pituitary gland was lodged); I.A.M., internal auditory meatus (for passage of the seventh and eighth cranial nerves into the ear region); O.T., olfactory tubercle on the cerebrum (the olfactory bulbs are not present on this cast but are approximately indicated by broken lines at the anterior end of the cast); P.L., pyriform lobe (or a lateral expansion of the forebrain); S.S.S., suprasylvian sulcus (representing another infolding of the brain).

Aires specimen, and also in this as far as can be judged from the skull, the olfactory bulbs are comparable to *Phenacodus* in bulk, but longer, narrower, less divergent, and with more slender peduncles. The olfactory tubercles are relatively even larger and more definite than in *Phenacodus*. The pyriform lobes are equally large and prominent, but this is due rather to their expansion vertically and posteriorly than to their dorsal exposure, or lateral expansion, for in dorsal view they are more limited than in *Phenacodus*, and chiefly posterolateral and posterior rather than definitely lateral to the neopallium.

The neopallium is even more distinctly triangular than in *Phenacodus*, with the anterior end much narrower than the posterior. The longitudinal scissure is well defined, but not so abnormally developed as in *Phenacodus*, and a vague trace of the venous sinus, at its posterior division point, is visible. The nearly straight and horizontal rhinal fissure is distinctly recognizable. The region of the fossa sylvii is a large ventrolateral depression, and from it a broad, short fissure extends upward and backward. Still more posteriorly, on the posterior part of the neopallium, is a broad and vague depression or sulcus apparently running forward and inward to or toward the upper end of the fissura sylvii, but the details are not impressed on the bone with sufficient definition to determine their exact relationships. On the present specimen there is very vaguely visible but on the Buenos Aires specimen quite distinctly developed, a simple, sagittal sulcus lateralis. The neopallium thus has two longitudinal dorsal gyri, marginal and suprasylvian, and distinct frontal and temporal lobes, the latter larger.

In the Buenos Aires specimen a large blood-vessel follows approximately the course of the rhinal fissure. What may be the homologous vessel in this specimen seems to be faintly indicated at the posterolateral corner of the cerebrum, immediately below the rhinal fissure. A small vessel ascends the anterior surface of the temporal lobe, just behind the fissura sylvii.

A round median swelling on the ventral surface, between and slightly posterior to the pyriform lobes, indicates the fossa hypophyseos, which is considerably more definite than in *Phenacodus*.

Laterally the flocculi and pyriform lobes are only about 2 mm apart, but dorsally, between the anterior end of the vermis and a line tangent to the posterior ends of the temporal lobes, about 6.5 mm intervenes between cerebrum and cerebellum. In this space is a broad, roughly rectangular, almost flat, depressed area. Here the midbrain must have been exposed, and the corpora quadrigemina were probably clearly visible in the dorsal view of the original brain, but they have left no impression on the bones. Probably part of the cerebellum underlay this area also, but in this specimen, not yet fully adult, it had not yet modeled the overlying bone.

The cerebellum is clearly divided into a large vermis and relatively very small hemispheres. The vermis is divided by two real but indefinite transverse sulci into three lobes. These are probably the true anterior, median (simplex), and posterior (posteromedian) lobes and their divisions the fissura prima and fissura secunda, but this is not absolutely certain, as these primary cerebellar fissures are not invariably more prominent than secondary sulci. The anterior lobe is also featureless (in the cast) confined to the dorsal surface, and is transverse, about half as long as the anterior lobe. The posterior lobe has about the same length as the anterior, but is less even, with vague traces of transverse sulci, and lies almost entirely on the more vertical occipital exposure of the cerebellum.

The cerebellar hemispheres are very vaguely divided into anterior and posterior moieties on the dorsal and occipital surfaces. Below the anterior part and extending still farther forward is the strongly developed "flocculus," elongated anteroposteriorly. From the posterior lobe of the cerebellar hemisphere, a sharp crest runs downward and backward, bounding the posterior end of the petrosal fossa.

The nerve exits are all well shown. The optic chiasma and optic canals occupy the same positions as in *Phenacodus* and the strong canal for III, IV, V_{1-2}, and VI, leading to the anterior lacerate foramen, is also very similar although relatively slightly closer to the midline. V_3 also has the same position and relationships as in *Phenacodus*. The large internal auditory meatus (VII, VIII) is ventral to and in part slightly posterior to the flocculus, near the center of the well defined periotic fossa. Posterior and slightly ventral to it, at the lower end of the post-periotic crest from the cerebellar hemisphere, is the relatively small posterior lacerate foramen (IX, X, XI), quite as in *Phenacodus*; XII also is as in *Phenacodus*, except that it has a second smaller and more dorsal root.

In or near the fossa sylvii there appears to be a small vascular foramen, not very clearly or certainly shown because of slight breakage in this region. The strong entocarotid vessels seen in *Phenacodus* are here absent, doubtless displaced by the development of the bullae. There is a pair of small vascular foramina in the basioccipital near the midline, slightly anterior to the level of the posterior lacerate foramen.

The most prominent vascular opening is directed backward from the posteroexternal angle of the cerebrum, just below the rhinal fissure. I was at first inclined to consider this an artifact, but its presence and clearly natural character in the Santa Cruz typotheres confirm its reality as an original endocranial feature. From it comes the great blood sinus so prominent near or on the rhinal fissure in the Buenos Aires specimen and also present, but less obvious, on the present cast. This prominent venous foramen apparently communicates with the epitympanic sinus, the postglenoid foramen, or both. Its relationships may be cleared up more fully by a detailed study of early notoungulate skull structure now in progress. It seems to be an important and unusual character.

Despite the inevitable differences of preservation, the brains of *Phenacodus* and *Notostylops* are seen to be remarkably similar. The general proportions, degree of development, and many details are closely comparable and the differences are mostly minor and not greater than may commonly be seen between the braincasts of members of a single family. While the skull of *Notostylops*, and its teeth, have been rather strongly modified, the brain is conservative and clearly has progressed hardly at all beyond the primitive ungulate type, apparently also retaining evidence of some special affinity with *Phenacodus*, or with the Condylarthra in general.

Among the numerous details in which *Notostylops* differs from *Phenacodus*, the following seem to be the most definite and important:

(1) braincast slightly longer in proportion to its width;
(2) olfactory bulbs of different shape (see above);
(3) pyriform lobes extending somewhat more posteriorly and less laterally;
(4) neopallium more triangular in outline;
(5) gyri probably developed in much the same pattern and to about the same degree, but some difference in detail not excluded by the known material;
(6) fossa hypophyseos relatively deeper and more distinct;

(7)   cerebellum shorter relative to cerebrum;
(8)   pons underlain by two longitudinal ridges (vessels?);
(9)   hypoglossal canal double (at internal end);
(10) course of entocarotid different;
(11) large vascular opening at posterolateral angle of cerebrum.

Several of these characters, perhaps most of them, are due to or related to relatively superficial habitus changes, shape of skull, development of bullae, etc., rather than to any more deep-seated or phyletic distinction.

Length of skull (estimated, rostrum imperfect)	*ca.*	115 mm
Length of braincast exclusive of medulla (estimated, olfactory bulbs absent)	*ca.*	55 mm
Length of olfactory bulbs (estimated)	*ca.*	12 mm
Length of cerebral hemispheres		23 mm
Width across cerebral hemispheres (pyriform lobes)		30 mm
Length of cerebellum (vermis)		13 mm
Width of cerebellum (flocculi)		23 mm

Source of Text

(1933) Braincasts of *Phenacodus, Notostylops* and *Rhyphodon, Amer. Mus. Novitates* No. 622, pp. 1–19.

Sources of Other Illustrations

(1937) Skull Structure of the Multituberculata, *Bull. Amer. Mus. Nat. Hist.*, Vol. 73, pp. 727–763. (Source of Fig. 2–3.)
(1933) The Ear Region and Foramina of the Cynodont Skull, *Amer. Jour. Sci.*, Series 5, Vol. 26, pp. 285–294. (Source of Fig. 2–4).

Other Relevant Reference

JERISON, H.J. (1973) *Evolution of the Brain and Intelligence*, Academia Press, New York and London. (Based primarily on cranial endocasts of fossils and with almost exhaustive bibliography of other publications about them.)

A Way to Look Inside Fossils

Comment

All organisms have internal anatomical features that are just as important as and often more important than their external ones. When dealing with recent organisms the methods for internal study are well known, standardized techniques which may be quite simple or which may have requirements for costly apparatus and for special skills but which are still standardized and involve widely marketed apparatus and widely taught skills. Among the most useful of these methods for recent organisms is serial sectioning, involving preparation such as by decalcifying or staining, followed by embedding and then slicing by a microtome and mounting on slides.

Many fossils also have important internal structures that cannot be adequately prepared for study by gross dissection and that can best be studied by serial sections. Among many examples are the internal arm-bearing structures (brachidia) or the hinge teeth of fossil brachiopods and the internal ear structures of fossil mammal skulls. It is rarely both practical and useful to prepare such sections as thin slices. The alternative is to grind the specimens a little at a time and to record the cross-sections thus successively exposed. A rather complex device for that purpose was invented by the British paleontologist W.J. Sollas as early as 1903 and later used by him and I.B.J. Sollas in the study of a fossil reptile skull. Since then a considerable number of other apparatuses for that purpose have been made and used in many different studies, perhaps the most elaborate and outstanding the remarkable and voluminous researches on early (Paleozoic) fossil fishes and other fish-like vertebrates by Erik A:son Stensiö and his students and colleagues in Stockholm.

Although one such device was for a time commercially produced, most of them have been one of a kind, made by hand in an institutional workshop and hence costly in time, labor, and money, which involves a serious problem. The problem arose for me when I was studying South American fossil ungulates. The most important group of those, the order Notoungulata, is characterized by an unusual and complex ear structure. Although this could be studied in part by its features external on the skull and in part by preparation of specimens broken through the ear region, it became clear that really thorough study would require at least one complete sequence of serial sections ground at close intervals. The funds and workshop skills for duplicating or purchasing any previously described apparatus or for constructing a new one of equal complexity were not available. It therefore was necessary to devise a new apparatus capable of the necessary precision but with cash cost near zero and time and labor minimal. The result was described in the paper reprinted below with slight deletions and one small verbal change. The specimen shown in the apparatus in Fig. 2–5 was the cranium of a small notoungulate, which was later described in detail in another paper cited below.

Although a description of a serial grinding technique had been in print for thirty years when I wrote this paper in 1933, the technique still had been little used. Since then it has been widely used in both vertebrate and invertebrate paleozoology, although as far as I know my apparatus has not been much copied, perhaps because more elaborate ones have become more readily obtainable. Since 1933 a useful adjunct to serial grinding has come into use: the preservation of each ground section in a peel. Collodion peels were in use by paleobotanists early in the 20th century, but not for serial ground sections. I do not know just when the latter technique was first adopted, but it was in use at the Museum of Comparative Zoology in Cambridge, Massachusetts, at least as early as the late 1950s. This technique involves lightly etching each successive ground section and then applying a cellulose acetate sheet or a parlodion solution which when dry can be peeled off embodying the surface layer of the section. The structure embedded in the peels can then be studied and permanently preserved.

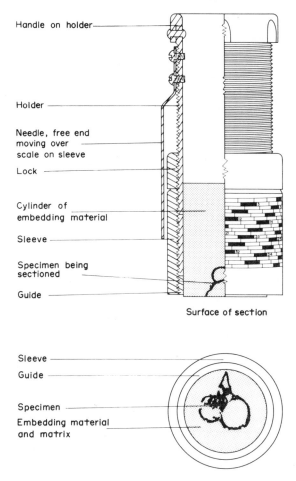

Handle on holder

Holder

Needle, free end
moving over
scale on sleeve

Lock

Cylinder of
embedding material

Sleeve

Specimen being
sectioned

Guide

Surface of section

Sleeve

Guide

Specimen

Embedding material
and matrix

FIG. 2–5. The serial sectioning apparatus described in the text. Above, side view, left half shown in vertical median section and right half in external view. Below, working face with a ground section of an imbedded specimen (cranium of a notoungulate) in place.

Text

Thin sections are extensively used in the study of fossils. They are part of the standard technique for the study of some invertebrates (e.g. fusilinids) and plants (silicified wood, coal balls, etc.) and they are occasionally employed in vertebrate paleontology (e.g. the work by Carter on enamel histology and that by Moodie on pathologic tissues and ossified tendons). Serial thin sections, as employed so extensively in recent biology, would be of still greater value. If, however, such series were to be made with paleontological materials, the space lost between each section would be many times greater than the thickness of the sections themselves, and the successful preparation of many successive thin sections would be extremely arduous, require a high degree of skill and elaborate equipment, and involve at best some failures and many imperfect sections.

Serial thin sections are very seldom practical in paleontological research, and serial grinding must be used.

This technique is so promising and its results so remarkable that it is surprising to see how little use has been made of it. This relative neglect is apparently due mainly to two factors. In the first place, the most puzzling specimens and those most imperatively needing such a method for their elucidation are usually rare or even unique, so that any mode of study that destroys the original is unwarranted. This is an inherent disadvantage which sharply limits the use of ground serial sections, but it still leaves a large field of usefulness. Every collection does contain specimens that could be more usefully employed in this way than in any other. Furthermore, many symmetrical specimens, even though rare, could legitimately be divided into halves, one side to be used for sectioning and the other for surface morphology and permanent record. The second difficulty, which has probably been the more important reason for the neglect of this method, is that the preparation of adequate sections has usually demanded either remarkable skill and hard labor or else very expensive and complex apparatus.

The simple procedure here suggested seems in large measure to do away with the difficulty and expense of the serial sectioning of fossils. The results are valuable out of all proportion to the cost, in time or money, and it is hoped that a wide use of serial sections may result. The necessary apparatus need not cost more than a dollar or two and the procedure is very rapid and requires no special skill or training. The machines illustrated by Sollas probably give slightly more exact results, but in practice the present method proves to be entirely adequate for any reasonable purpose. More elaborate equipment or procedures could hardly produce a useful improvement in results.

This method involves no particular originality, and its publication is prompted only by the desire to share a useful tool with others. I am indebted to C.S. Williams for the actual construction of the apparatus and for some suggestions regarding its design and use. The method has been used in the study of parts of small mammal skulls. Serial sectioning does not appear to have been employed previously in the study of fossil mammals, but the results, which will be published elsewhere, are excellent and could have been obtained in no other way. Any paleontological specimens or parts of specimens could be used up to a diameter of perhaps 60 mm. Larger specimens would require an apparatus similar to that of Professor Sollas, or would have first to be cut into smaller segments.

The requirements of successful serial sectioning are the means to:
(1)   grind smooth, plane surfaces without chipping;
(2)   orient these surfaces with respect to some selected axis;
(3)   keep successive sections parallel to each other;
(4)   maintain a known and uniform distance between sections;
(5)   preserve the unground part of the specimen undamaged until used.

The simplest possible procedure, holding the specimen by hand and grinding against a lap or stone, obviously cannot meet any of these conditions to the required degree and some sort of specimen holder is therefore a primary need. In the method here described, the holder, proper, is a piece of ordinary metal pipe with inside diameter several millimeters larger than the maximum dimension of the sections to be ground and the outer surface threaded for most or all of its length. In addition there is required a metal sleeve, shorter than the holder and longer than the total thickness to be covered by any one series of sections, with internal threads fitting the external threads of the holder. The sleeve should be fairly thick and of hard metal, and one end should be ground or machined to a plane surface at right angles to the axis of the cylinder.

These two pieces are all that is absolutely necessary, but several additions increase the ease of use and accuracy of results and are easily constructed. As a guide, a very short section of the same sort of threaded pipe as the holder may be permanently fixed at the machined end of the sleeve, so that its outer end is flush with and forms part of that plane surface. Some sort of grip or handle at one end of the holder will make it easier to screw this into the sleeve. A shorter, separate sleeve segment screwed onto the holder between its grip and the main sleeve forms a convenient means of locking sleeve and holder together at any desired point. Finally a pointer fixed to the holder and passing over a properly calibrated scale on the sleeve will make it easy to measure the distances between sections. The scale may be calculated from the known pitch of the threads and the outside circumference of the sleeve and then drawn on paper and shellacked to the sleeve.

For grinding, a lap may be used by simply bringing the holder to bear against its plane surface by hand. If a lap is not available—and this may be found as quick and easy in any case—two or three ordinary, flat, rectangular sharpening stones of different degrees of fineness will serve just as well. It will be found easiest to hold the specimen in its carrier stationary and upright and to move the stone across it. Both stones and section should be kept wet and washed as needed. The stones may be kept plane, if necessary, by rubbing them against each other occasionally. As the amount removed at each cut is small, hand grinding is easy and rapid, averaging five minutes or less for each section.

In using the apparatus, the specimen to be ground is embedded, with the desired orientation, in one end of a cylindrical block of suitable material. In practice, plaster mixed so as to set with fine grain and few bubbles, dried, and then impregnated with thin shellac has been found excellent for embedding. The cylinder should be cast in the specimen holder, well greased to facilitate removal. The specimen is embedded in one end of the cylindrical plaster block, and the other end should project far enough for firm attachment in the holder, as detailed below.

After removal from the mold and any further preparation (such as thorough shellacking), that end of the block not containing the specimen is inserted into the holder far enough to be held firmly, but leaving the part with the specimen freely projecting. It is fixed in this position. The holder and the embedding cylinder now projecting from it are then screwed into the sleeve far enough to bring the specimen to the far end of the latter, the end machined to a plane surface. The exact amount to be removed at the first cutting is then exposed by screwing the holder farther into the sleeve. After the first grinding, the holder is again screwed in for a determined distance and the operation repeated. Each section is ground flush with the machined end of the sleeve and guide.

The planes so ground are exactly parallel because each is determined by the machined end of the sleeve, which is exactly transverse to the axis along which the specimen is moved between cuts. The end of the sleeve and guide is so much harder than the embedded specimen that it is easy to stop grinding when it is reached. It wears slowly, but in a given case of a series of over fifty sections the total wear was less than 0.1 mm, or less than 0.002 mm per section, which is completely negligible. The interval between sections is easily measured or predetermined by the amount to which the holder is screwed into the sleeve, and by making this the same each time, the sections are kept exactly equidistant from each other.

Care should be taken to keep the threads well oiled and to prevent the embedding cylinder from adhering to the guide, if one is used.

Primary records taken are a drawing or a photograph of each section. Whether drawings or photographs are preferable will depend on the nature of the specimen and on personal judgment. If photographs are used, each should be developed, printed, and retouched before the next section is ground. The specimen is left in the apparatus for drawing or photographing, and the apparatus itself is a convenient holder and means of orientation for this purpose. For additional orientation, two small holes bored in the embedding medium (but not through the specimen) may be helpful.

If desired, wax models or copies on sheets of glass (both suggested and described by Sollas) may be prepared from the drawings or photographs for further study.

Source of text

(1933) A Simplified Serial Sectioning Technique for the Study of Fossils, *Amer. Mus. Novitates*, No. 634, pp. 1–6.

Other References

KUMMEL, B., and D. RAUP (1965) *Handbook of Paleontological Techniques*, W.H. Freeman, San Francisco and London. (Serial grinding, pp. 212–224; peel technique, pp. 224–232.)

SIMPSON, G.G. (1936) Structure of a Primitive Notoungulate Cranium, *Amer. Mus. Novitates*, No. 824, pp. 1–31. (A study made by use of the apparatus described above.)

A Way to Tell Crocodiles Apart

Comment

Vertebrate paleontology would be much simpler, but perhaps less interesting, if all its specimens were complete skeletons. In fact with the exception of a few deposits, mostly of fossil fishes, whole skeletons are decidedly rare and the usual objects of study in this science are mere fragments and isolated pieces. Most mammals can be fairly well identified (not restored) from their dentitions alone or frequently even from single teeth. That is usually more difficult and sometimes virtually impossible for many other groups of vertebrates. (Of course the question does not arise for the vast majority of birds or for other toothless vertebrates.)

Crocodilians used to be more diverse and more plentiful the world around than they are today. They are still numerous in much of South America although less varied, less populous, and less wide-ranging now than a few million years ago. They comprise an important part of the faunal history on that continent, and while primarily hunting fossil mammals there I also found and collected various parts of fossil crocodilians. One find included most of the body skeleton but only one bone from the head, the right dentary or tooth-bearing bone of the lower jaw with all the tooth sockets but only two of the teeth. It turned out that identification of the creature and judgment as to its relationships would have to depend almost entirely on the tooth sockets in that bone, a tough-looking problem but one that turned out to be reasonably soluble. The specimen turned out to represent a definable new genus and species, which was named *Necrosuchus ionensis*, and to be related to a previously known North American genus named *Leidyosuchus*. The comparative method used in reaching these conclusions is discussed and illustrated in the following excerpt, here modified only by elimination of three footnotes not necessary in the present connection.

Text

The crocodiles are, in general, homodont animals. It is, however, well known that within this group there may be considerable difference in form between the various teeth of one individual and that even when all the teeth have approximately the same form they may differ markedly in size. Such tooth differentiation is particularly striking in various alligatorids, including the South American caiman-group, and it is also typical of *Leidyosuchus* and *Necrosuchus*. In attempting to classify *Necrosuchus ionensis* largely on the characters of its dentary, it was necessary to analyze this differentiation and to estimate its significance. The methods and results are here given from the point of view of the study of the affinities of this single species, but these results suggest that tooth differentiation among crocodiles is more exactly measurable and more significant than has hitherto appeared and that it will be worthwhile to employ similar methods more widely in the study of this group.

Tooth differentiation in form is seen in its extreme among Crocodilia in *Allognathosuchus*, with fairly typical crocodilian teeth (but highly differentiated in size) in the front part of the jaw and depressed, crushing teeth (not much differentiated in size) in the posterior part. Although aberrantly strong in this genus, such qualitative differentiation is characteristic of the Alligatoridae and occurs in great or small degree in most genera of that family. It is nearly or quite absent in *Leidyosuchus*, *Necrosuchus*, and in the Crocodilidae generally.

On the other hand, *Leidyosuchus* and *Necrosuchus* have quantitative differentiation almost equal in degree to that of the jacaré-like alligatorids and similar, but not the same, in kind. This pattern of quantitative differentiation is clearly shown in the accompanying graphs (Figs. 2–6, 2–7), in which the serial numbers of the teeth are used as abscissas and the anteroposterior diameters of their alveolar mouths as ordinates, with a line joining the points so determined giving a pattern for each single specimen.

The first noteworthy fact is that these patterns are characteristic for the specimens here treated and may be inferred to be so for the species. Each has a distinctive pattern and the obvious resemblances and differences between these patterns correspond in an unexpectedly clear and exact way with the views as to classification and affinities based on other criteria. Thus specimens of the three species already referred to *Jacaré* (by Patterson, who calls the genus *Caiman*) on other grounds obviously give more variants of the same pattern, which is hence (subject to more extended investigation) inferred to be generic. The type of *Eocaiman cavernensis* differs more from these three species than they do among themselves, but has some basic resemblance to them and resembles

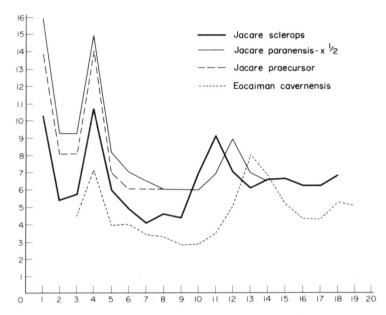

FIG. 2–6. Dimensions of anteroposterior diameters of sockets of lower teeth in South American alligator-like crocodilians. *Jacré* is a living genus, *J. sclerops* one of its living species and *J. paranensis* and *J. praecursor* extinct species. *Eocaiman* is a much older extinct genus. *J. paranensis* is about twice as large as the other species and its measurements are divided by two to facilitate comparisons. Otherwise the measurements are in millimeters. The teeth are numbered from front to back.

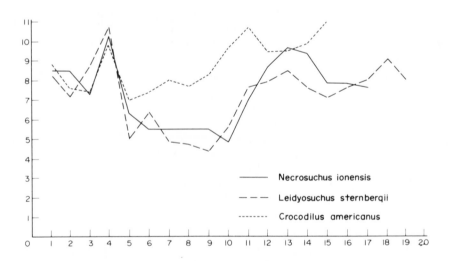

FIG. 2–7. Dimensions of anteroposterior diameters of sockets of lower teeth of crocodile-like crocodilians, graphed in the same way as Fig. 2–6. *Crocodilus americanus* is the living Florida crocodile. *Leidyosuchus sternbergii* is an extinct North American Paleocene crocodile. *Necrosuchus ionensis* is a similar extinct South American Paleocene crocodile. The graph supports evidence of its affinities with *Leidyosuchus*.

them more than it does *Crocodilus americanus*, for instance, which is in agreement with the conclusion reached on other criteria that it belongs to a distinct genus of the *Jacaré* group.

In the other series of graphs (Fig. 1–7), *Necrosuchus ionensis* is shown to give only a slight variant of the *Leidyosuchus sternbergii* pattern, and this sort of pattern is seen to be unlike that of the jacarés and also unlike that of a typical true crocodile, *Crocodilus americanus*.

The most striking mandibular tooth size characters of the *Jacaré* group (including *Caiman*) seem to be:

(1) first and fourth teeth subequal, larger than any others in the jaw with occasionally a single exception;

(2) second and third teeth markedly smaller than first or fourth, usually subequal [(1) and (2) were also true of *Eocaiman cavernensis* although the first two alveoli do not permit exact measurement];

(3) fourth tooth followed by a series of six to eight smaller teeth, the smallest tooth in the jaw occurring in this series;

(4) a single tooth, eleventh to thirteenth in various cases, enlarged, nearly or quite as large as the fourth, with the series becoming abruptly smaller both anterior and posterior to it;

(5) size differentiation as a whole strongly developed;

(6) total number of teeth 18 to 22.

The characters of *Necrosuchus* and *Leidyosuchus* are:

(1) first tooth smaller than fourth; fourth the largest in the jaw but first equalled or exceeded by the thirteenth and sometimes by others;

(2) second and third teeth slightly unequal, one of them about equal in size to the first;

(3) a series of smaller teeth including among them the smallest in the jaw, following the fourth, as in the *Jacaré*-like genera;

(4) thirteenth tooth (exact position perhaps variable with more material) enlarged and nearly equal to fourth, approximately as in the *Jacaré* group, but transition to this tooth somewhat less abrupt anteriorly and markedly less so posteriorly, some of the more posterior teeth being nearly or quite equal to the thirteenth in size;

(5) size differentiation well developed, but somewhat less than in the *Jacaré* group;

(6) total number of teeth in early species 18 or 19, within the *Jacaré* group range (but number considerably greater in the Paleocene species of *Leidyosuchus*).

These differences from the jacarés are, on the whole, points of resemblance to *Crocodilus* and closely allied forms. The corresponding characters of *Crocodilus* appear to be as follows, subject to emendation since I have made no attempt to measure and plot a majority of the species of this protean genus:

(1) first tooth equal to or slightly smaller than fourth; the fourth usually equalled by two or several posterior teeth;

(2) second and thrid teeth subequal, usually slightly smaller than first;

(3) fifth tooth abruptly smaller than fourth and either the smallest in the jaw or about equal to the third; after the fifth the series increases steadily in size to about the tenth or eleventh and there is no markedly differentiated small series as in the preceding two groups;

(4) teeth from about the eleventh to the end of the series subequal and nearly or quite equal to the fourth; there is often a tendency for the eleventh tooth to be somewhat the largest;

(5) size differentiation markedly less than in either of the preceding two groups;

(6) total number of teeth usually 15.

These graphs show the kind or pattern of size differentiation. It is more difficult to measure and to compare the degree of such differentiation in any adequate way. For this purpose a coefficient of differentiation is tentatively proposed that seems to fill the need for the particular groups here discussed. It is taken as one hundred times the mean deviation divided by the arithmetic mean of the anteroposterior diameters of the first fourteen mandibular alveoli for each individual. The mean deviation for all the teeth is a measure of average size differences, hence of size differentiation, but is not comparable from one species or specimen to another because it depends also on absolute size and on the number of teeth, which are foreign to the character for which a measure is sought. These are eliminated by limiting the alveoli involved to the first fourteen and by dividing by the mean. In the groups of immediate interest here, the size differentiation is largely in the teeth anterior to the fifteenth. The number of rather uniform teeth posterior to the fourteenth varies from one in *Crocodilus* to fourteen in *Leidyosuchus multidentatus*. Obviously if all the teeth were included in the coefficient the figures for these two would not be comparable: that for *Crocodilus* (all species) would be relatively too high and for *Leidyosuchus multidentatus* (and to less extent other species of that genus) relatively too low. Basing the coefficient only on the anterior teeth, where the principal differentiation does occur, is a valid and simple way to avoid this difficulty, although not a perfect solution. Dividing by the mean makes the figure independent of the absolute size and permits valid comparison between individuals and species of different sizes. For instance, *Jacaré paranensis* is clearly a typical *Jacaré* as regards size differentiation, but the mean deviation for the first fourteen alveoli in an individual is 4.7 mm as compared with only 1.6 mm for an individual of *Jacaré sclerops*: the figures are not comparable because *J. paranensis* is much larger. Their coefficients are nearly the same.

Coefficients thus calculated for some of the species here mentioned are:

Jacaré paranensis	27.7
Jacaré sclerops	24.6
Leidyosuchus sternbergi	22.3
Necrosuchus ionensis	21.5
Crocodilus americanus	11.9

Source of Text

(1937) An Ancient Eusuchian Crocodile from Patagonia, *Amer. Mus. Novitates*, No. 965, pp. 1–20.

Ratio Diagrams

Comment

Related organisms commonly differ not only in size but also in proportions, and this must be kept in mind when comparing them. The comparison must involve not only the absolute measurements of various dimensions but also the ratios between measurements of the same dimensions in two or more samples of populations being compared.

That necessity strongly impressed me around 1940 when I undertook a revision of all the large Ice Age (Pleistocene) fossil cats that had been found in North America. Although only those the size of the living pumas or larger were included, they differed markedly both in size and in proportions and a considerable number of different generic and specific names had been given to them. It was necessary to use ratios in evaluating the differences, and that suggested a great many calculations and serious difficulties in visualizing them overall and judging their significance. As I struggled with those problems I had an idea that really was quite simple but that in retrospect does, I think, merit being called an inspiration. As far as I know the idea had never occurred to anyone before, and certainly it was useful because it has been quite widely used ever since it was published.

Publication was in an appendix to the paper on the large Pleistocene cats. That appendix is here reprinted, somewhat modified and slightly shortened by changing or omitting parts referring most specifically to the cats and therefore with less direct application in the wider usage foreseen.

Text

Several figures of this paper (e.g. Fig. 2–8) are constructed on a principle that I have not seen used elsewhere and they require some explanation. The method is one that lends itself readily to several useful types of graphic analysis and comparison and may therefore find wide application. I have found it convenient as an aid in distinguishing species and determining affinities, in sorting collections of bones, and in various ways beyond those illustrated in this publication.

The basic purpose of the diagram is to represent each of a number of analogous observations by a single entry and to plot them in such a way that the horizontal distance between any two of them will represent the ratio of either one of those two to the other. A simple plotting of calculated ratios has various and valuable properties, but it does not have the basic property sought for these more generalized ratio diagrams. It shows the ratios of various observations to one fixed standard or between fixed single items in one series to similar single items paired with these in a related series, but it cannot show ratios between any two observations among many.

Consideration of the desired properties shows that the scale used must be logarithmic and not arithmetic. For instance, given absolute values $a = 1$, $b = 2$, and $c = 4$, the distance plotted between a and b should be the same as between b and c, because $a:b = b:c$. This is true on logarithmic but not on arithmetic coordinates. Since it is desired to ignore absolute values and represent only ratios, the simplest approach is to plot the logarithms of ratios. The logarithm of a ratio is the difference between the logarithms of the two

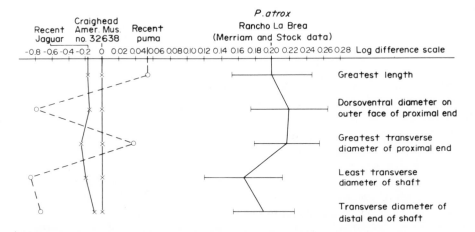

FIG. 2–8. Ratio diagram of dimensions of a second metatarsal (a hind foot bone) of various large felines. The Craighead specimen is a fossil from a cave in Tennessee, a region where jaguars are not known to have been present since the Ice Age (Pleistocene). The graph shows that the specimen is slightly larger than the recent jaguar being compared, but as the lines are almost parallel in the graph the specimens have almost exactly the same proportions and the fossil is in fact a jaguar. The puma bone is near the size of those of the jaguars but has quite different proportions. *P.* (for *Panthera*) *atrox* is an extinct lion best known from Rancho La Brea in Hollywood. The horizontal lines show the observed range in a large sample from Rancho La Brea described by Merriam and Stock and the ratio line is drawn through the mean values. The size is much larger throughout than jaguars and pumas. The proportions are different from either but more like the jaguars.

absolute measurements entering into the ratio. The easiest method, then, is to plot the differences between logarithms. By using these differences as they are, not converting them to antilogs, a step may be saved and also ordinary arithmetic graph paper may be used, since plotting logs on this gives the same result as plotting antilogs on logarithmic paper.

For calculation, the direct measurements are first converted to their logarithms, three decimal places generally sufficing. Some one observation is then taken as "standard," to represent zero difference in logarithms which corresponds with the ratio 1.00. Observations larger than this then fall to the right of it at distances determined by their ratios to it, and smaller observations similarly fall to the left. Although the differences are thus calculated from some one standard, the resulting diagram shows not only ratios to that standard but also ratios of any combinations of observations: once the diagram is made, the zero point, or ratio 1.00 point, may be placed anywhere and ratios of all other observations to that point will still be correctly represented.

The arithmetic involved is much simpler than calculating even one set of ratios, not to speak of all possible sets, in spite of the fact that the resulting diagram does represent all possible sets. For instance, all the arithmetical calculation for one example is as shown in Table 2–1.

These figures can then be plotted against the log difference scale as shown.

A scale for reading ratios directly from the diagram can easily be constructed. By copying this on a separate slip of paper, a movable scale can be made and the diagrams have the property that if 1.00 on the ratio scale be set at any specimen (whether the standard or not), the values of the ratios of all other specimens (set on the same horizontal) to this one can at once be read on the scale.

TABLE 2–1

(Variate: length of second metacarpal in three cats)

Specimen	Measurement (mm)	Log	Difference from log of standard of comparison
U.S.N.M. No. 12840	94	1.973	0 (The standard here used)
Recent jaguar	68.3	1.834	− 0.139
Recent puma	77.2	1.888	− 0.085
P. atrox from	maximum 124.4	2.095	+ 0.122
Rancho La Brea	app. mean 2.048		+ 0.075
	minimum 100.4	2.002	+ 0.029

In study it is convenient to adhere to a single scale as far as possible. On study sheets, using arithmetic graph paper ruled in millimeters, I have found scaling log difference 0.02 as 10 millimeters to be most convenient.

Points marked on a single horizontal line represent different values of one variate. In the diagrams in this paper they represent homologous dimensions of different specimens, but any series of values of one variate can be used. The further and perhaps greatest usefulness of the method lies in the arrangement and interpretation of diagrams in which several different variates are involved, each represented by one horizontal series and these series arranged one below the other. The horizontal single variate series are so placed that related points fall into a single vertical line. For instance, different measurements on one individual may be placed in a vertical line or mean values for a unified sample may be so placed, or values for a group of specimens may be thus arranged in order to test the hypothesis that they represent one species and to see what comparable species they most nearly resemble in their ratios to each other.

The diagrams so constructed have many different uses and properties, some of which will be mentioned and others of which will be seen as the method is used.

If a series of measurements involves the same ratios as the standard of comparison, or, dealing with two individual animals, if a second animal has the same proportions throughout the parts measured as has the animal with which it is being compared, then this series being compared will also fall into a straight vertical line, regardless of whether the two animals are of the same absolute size or not. The more nearly the comparative series approaches the proportions of the standard series, the nearer will the comparative series come to falling into a vertical line.

If, as is generally the case, it can be postulated that scattered and isolated specimens belonged to animals with about the same proportions as some available standard of comparison, then the method makes it possible to compare the sizes of two or more animals known only by different specimens that are not homologous and that cannot be directed compared.

The method can also be used to estimate the relative size of missing parts of specimens. For example, if in a ratio diagram the available measurements of a specimen or species are nearly in a straight line, it is probable that a missing measurement would also be near that line.

Finally, when the sets of measurements used represent individuals or species that do, in fact, have different proportions, such diagrams show in a simple and immediately apparent way what all these differences are. For example a ratio diagram was made for nine different homologous measurements in two species of large cats, *Felis onca*, the living jaguar, and *Panthera atrox*, an extinct lion. To make a similar comparison by nongraphic means would have involved calculation of 108 ratios, and the results would not have been as clear and usuable as in one simple ratio diagram.

Source

(1941) Large Pleistocene Felines of North America, *Amer. Mus. Novitates*, No. 1136, pp.1–27.

The Concept and Definition of Homology

Comment

The terms "homology" and "analogy" were coined by pre-evolutionary comparative anatomists. Concepts connected with those terms later became crucial in evolutionary biology, but there has not always been full agreement as to just what those concepts are. In the 1940s there was extensive dispute that is now almost but still not quite settled. Apart from more subtle differences in definitions, the major disagreement was between biologists who although themselves evolutionists wanted to retain these terms in their oldest senses as defined in nonevolutionary ways and those who discarded what they considered obsolete definitions and followed post-Darwinian evolutionary definitions and concepts.

The late Otto Haas, who had been a lawyer in pre-Nazi Austria but who changed from an amateur to an accomplished professional invertebrate paleontologist after coming to the United States, became much concerned with this partly linguistic, partly legalistic, and partly philosophical subject. He wrote a long manuscript discussing the history of these two terms and many others more or less related to them. He asked me to read and discuss his manuscript, which I did at such length that he insisted that I become a coauthor, which I consented to do but only as junior author. We agreed completely on almost all essential points, but did disagree on some minor matters of definition and categorization. Thus the part of the final publication here extracted for republication became to some extent a dialogue between the two authors. The excerpt is here slightly abridged and much of the whole paper is omitted.

There is now a strong consensus, if not absolutely complete agreement, in favor of the definitions and concepts that Haas and I endorsed.

Text

Homology (Owen, 1843, = special homology, Owen, 1847) is here defined as a similarity between parts, organs, or structures of different organisms, attributable to common ancestry.

It will be noticed that—despite Hubb's advocating "that we think and write of homologous functions in the same way that homologous structures have been treated"—functions are not included, in the above definition. It is believed that the function of organs may or may not be the same in different organisms, but that functions *as such* are never homologous the one to the other. What Hubbs really seems to have in mind is to use, in certain cases, identity of function as a criterion of homology. There is certainly no objection against this, but it must be kept in mind that identity of function has been considered, from the very beginning of the terminology under discussion, to be *the* characteristic of analogy rather than homology.

Note by Simpson: The preceding paragraph represents the view of Haas. I do not quite agree, although the disagreement is of slight importance and is in part merely a matter of verbal expression. Hubb's point was to stress the difficulty and, at times, the undesirability of considering structure and function separately. Similarity of function is not exclusively characteristic of analogy. It is often, indeed usually, characteristic of homology, also. In fact, from one point of view, analogy and homology are two alternative theoretical explanations of similarity both in function and in structure. In such cases, I can see no objection to speaking of homologous functions, as an expression of opinion as to the origin of a dynamic complex including the functions along with the structures performing them. That "functions *as such* are never homologous the one to the other" is a matter of definition, only. I agree that such a definition is desirable if phrased that functions, considered as abstractions and without consideration for the structures that perform these functions, should not be spoken of as homologous, but that homology frequently involves both function and structure and in this combined sense it is proper to speak of the homologous functions performed by homologous structures. It is also true, but less common, that the functions of homologous structures may not be homologous. [End of note by Simpson.]

Before attempting definitions of either analogy or homoplasy or both, it will have to be examined whether or not these two terms are synonymous. This question has vexed students ever since homoplasy, the younger term, was proposed by Lankester in 1870. Lankester, himself, answered the question as follows:

> It may be said that the term "analogy," already in use, is sufficient to indicate what is here termed "homoplasy"; but analogy has a wider signification given to it in which it is found very useful to employ it, and it could not be used with any accuracy in place of homoplasy. *Any* two organs having the same function are analogous, whether closely resembling each other in their structure and relation to other parts or not; and it is well to retain the word in that wide sense.

Hubbs arrives at the conclusion that Lankester "merely substituted 'homogeny' for special homology and 'homoplasy' for analogy." Thus he equates homoplasy with analogy.

In our opinion, however, analogy and homoplasy are not synonyms.

If "analogy" is to retain, beyond its rather vague general significance, any technical meaning in comparative anatomy and phylogeny, it can be but the one implying function or use. This is the meaning given to this term by Owen in 1843 and maintained in the biological sciences, though with certain qualifications, ever since. Even Hubbs seems to agree with Boyden in considering "agreement *in use* [italics ours] of characters (structures or functions)" of distinct evolutionary orign as essential for "a modern concept of analogy."

If use or function is implied by the term analogy, then this term does not connote, as do both homology and homoplasy, a morphological resemblance. Theoretically at least, parts, organs, or structures of two or more organisms may serve the same use or have the same function without being similar. On the other hand, no meaning as to use or function is implied in the term homoplasy. It might be rather tempting to have to do with one dichotomy only: homology (similarity due to common ancestry)—analogy (similarity not due to common ancestry). However, after clarification of the concepts involved the need for such a simple dichotomy seems to be fulfilled better by the one contrasting homologous and homoplastic similarities, analogy being on a different categorical level from homology and homoplasy. Furthermore, homoplasy is, at least primarily, an evolutionary process bringing about homoplastic similarities, whereas analogy is from the outset a relation between two organisms or parts thereof.

Thus, homology and analogy are not mutually exclusive, and Boyden's examples of homologues which are at the same time analogous, or of nonhomologues which are not analogous all the same, are still considered valid.

Note by Simpson: This involves the second of the points on which Haas and I are not in complete agreement, although, again, our disagreement is not fundamental and does not affect the more important conclusions such as the definitions of homology and homoplasy.

Homology, as we agree, is best defined as similarity interpreted as due to common ancestry. Homoplasy, as we also agree, is best defined as similarity (or as including *any* process leading to similarity) that is not explicitly interpreted as due to common ancestry. Both terms rather than being purely descriptive (as in Boyden's definition of homology, for instance) express an opinion, one positive and one negative. Homology expresses an opinion as to how the similarity arose. Homoplasy expresses an opinion as to how the similarity *did not* arise, i.e., that it did not arise by homology, but it does not express an opinion as to how the similarity *did* arise.

I do not, in quite the same way as Haas, see these as alternatives at the same categorical level. The set is not positive, "a" and "b" as mutually exclusive categories, but is a dichotomy of "a" and "not-a." Under "not-a" it is still possible to have a sequence of alternatives, such as "b," "c," etc., that are positive categories on the same level as "a." Quite clearly by a few authors and vaguely by many, "analogy" has been used as an alternative of type "b," although, as the quotations gathered by Haas show, it has often been used for "not-a" and hence as a synonym of homoplasy.

Haas and I agree that equation of analogy with homoplasy is undesirable, despite the considerable weight of usage, because it omits the element of function, which historically belongs in the concept of analogy and which still tinges, at least, that concept in the minds of most users of the term. I feel, however, that similarity in structure along with similarity in function is equally an essential part of the concept of analogy, as that term has almost always been used. Moreover, the implication is usually present to some degree and it has sometimes been explicitly stated that the structural similarity here in question is not due to homology but is correlated with community of function as opposed to community of ancestry. It is in this sense that analogy is a true alternative (but not the only alternative) to homology as a positive category on the same level, a "b" category rather than a "not-a" category or something on a different level altogether. That is, analogy, when used in this way, expresses a positive opinion, or theory, that a structural resemblance is correlated with function, just as homology expresses the view that it is correlated with common ancestry.

Unlike homoplasy, analogy offers an alternative theory as to the basis of the resemblance in question. It is in one way a subdivision of homoplasy, designating, like homoplasy, a similarity that *is not* interpreted as homology but adding to this purely negative definition of homoplasy the opinion that the similarity *is* to be interpreted in another way. Although parallelism and convergence may also correctly be called subdivisions of homoplasy, this does not place them on a level with analogy. They belong in a different categorical system because they are (as here defined) descriptive only, while analogy and homoplasy, itself (but only negatively), are interpretive. [End of note by Simpson.]

Source

HAAS, O., and G.G. SIMPSON (1946) Analysis of Some Phylogenetic Terms, with Attempts at Redefinition, *Proc. Amer. Phil. Soc., Vol. 90, pp. 319–349.*

An Exercise in Homology

Comment

Almost all studies of the relationships and evolutionary histories of organisms involve the concept of homology and the recognition of homologous anatomical parts. In particular instances that may be so evident as not to require explicit notice. In other instances the recognition of homologies may be a disputed problem and its solution may determine what the affinities between groups of organisms are and how those groups are classified. The concrete example here given to illustrate the application of the concept of homology is of the latter sort.

The groups now generally classified as the orders Pantodonta and Dinocerata include large extinct hoofed mammals that were abundant in North America and Eurasia during the early part of the Cenozoic Era, the Age of Mammals. In the late 19th and early 20th centuries the two groups were usually considered closely related and often united in a single order called Amblypoda. The arguments about their affinities and classification finally rested largely on different views as to the homologies of cusps and crests in their molar teeth. The discussion of that topic reprinted here arose in connection with the description of a new genus, *Probathyopsis*, belonging to the Dinocerata.

The teeth of mammals are now almost always described in terms of a special nomenclature started by H.F. Osborn in 1888 and since then much augmented. The example here reprinted involves the identification of the homologies of cusps expressed in this nomenclature and shown in the figures (2–9 to 2–14) by standard abbreviations for their names. The cusps involved and their abbreviations are as follows:

Upper molars		Lower molars	
Hypocone	Hy	Entoconid	En^d
Metacone	Me	Hypoconid	Hy^d
Metastyle	Ms	Hypoconulid	Hl^d
Paracone	Pa	Metaconid	Me^d
Protocone	Pr	Metastylid	Ms^d
Protoconule	Pl	Paraconid	Pa^d
Parastyle	Ps	Protoconid	Pr^d

The suggested homologies are clear enough from the illustrations without necessarily knowing the cusp nomenclature.

Of the genera named in the discussion *Probathyopsis*, *Prodinoceras*, *Eobasileus*, and *Uintatherium* are now classified as Dinocerata and *Pantolambda* and *Coryphodon* as Pantodonta. The evidence presented in 1929 and here reprinted was largely instrumental in eventual recognition that these genera belong in two quite distinct orders of mammals.

Text

Both Matthew and Wood considered the present genus [*Probathyopsis*], which they mentioned but did not describe, as increasing the probability that the Coryphodontidae and Uintatheriidae are parallel phyla, quite distinct from each other. The question of the relationship between these two groups is too complex for full discussion here, but the bearing of *Probathyopsis* and *Prodinoceras* on this problem must be mentioned.

Osborn in 1898 elaborated the conception of the Pantolambdidae, Coryphodontidae, and Uintatheriidae as successive families forming a structually ancestral series. According to his view, the *Coryphodon* upper molar arose from one like *Pantolambda* by the rotation of the ectoloph and reduction of the paracone, and from this the *Uintatherium* molar was derived by further rotation of the ectoloph, so that the metacone became approximated to the protocone. The premolars of *Pantolambda* and *Coryphodon* are closely similar and have homologous parts. Those of *Uintatherium*, according to Osborn's theory of 1898, must have arisen by the complete suppression of the internal heel, and hence have had a very different history from the molars, which they closely resemble.

Wood has raised serious objections to this view. He holds that the *Coryphodon* and *Uintatherium* upper molar patterns are not formed by homologous elements, but that in the latter Osborn's "metacone" is the protocone, his "protocone" the protoconule, his "paracone" the metacone, and his "parastyle" the paracone; that is, that the *Uintatherium* molar has a normal trigon modified by an unusual type of lophiodonty.

Matthew advanced, only to reject it, the alternative hypothesis that in *Uintatherium* Osborn's "protocone" and "metacone" are the paracone and metacone, respectively, his "parastyle" and "paracone" the parastyle and metastyle, and that the protocone is lacking, but he concluded that the evidence of *Prodinoceras* (including the genus now named *Probathyopsis*) confirmed Wood's view. For *Coryphodon*, however, he advanced a new theory, namely, that the upper molars of this genus "apparently . . . are derived from something of the *Pantolambda* type, by uniting the anterior limbs of the paracone and protocone crescents into a single crest, and loss of their posterior limbs, the metacone crescent remaining little changed or losing its anterior[1] limb.''

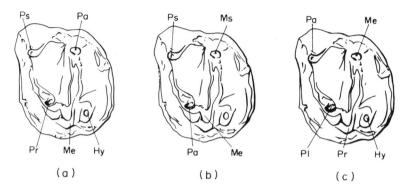

(a)   (b)   (c)

FIG. 2–9. Theories of cusp homologies in a left upper molar of *Uintatherium*, a North American Eocene member of the Dinocerata. The abbreviated symbols for the cusps are listed in the text. The three views as to possible homologies differ radically.

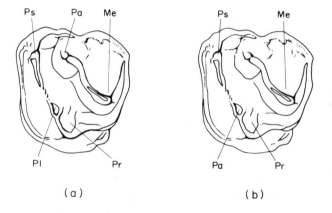

(a)   (b)

FIG. 2–10. Theories of cusp homologies in a left upper molar of *Coryphodon*, a North American Eocene member of the Pantodonta.

Prodinoceras and *Probathyopsis* seem to offer definitive evidence regarding the origin of the uintathere molar and to establish Wood's view beyond much question, as recognized by Matthew. The external ends of the two lophs are clearly the paracone and metacone, and the chief internal cusp is clearly the protocone. The so-called "protoloph" connects the paracone and protoconule, the "metaloph" ("ectoloph" of Osborn in 1898) unites the metacone and protocone. They are not homologous with any of the three perissodactyl lophs. There is an ectoloph in the Paleocene genera, particularly in the premolars, but in the later genera the deepening of the notch between the paracone and metacone causes the disappearance of the ectoloph as such.

1. A *lapsus calami* for *"posterior"*?

This view has the added recommendation that it considers the posterior premolars and anterior molars, almost identical in form, save for the presence of a hypocone in the latter, as composed of homologous parts instead of assigning to them widely different histories.

Careful restudy of the *Pantolambda* and *Coryphodon* molars seems to confirm Osborn's views as to cusp homologies in the latter, although not his mechanical interpretation of their origin. In *Pantolambda* the paracone and metacone are equal, strongly crescentic, their apices median. The last deciduous molar of Coryphodon gives a valuable clue to the origin of the pattern of its true molars. Its parts are clearly homologous with those of *Pantolambda*, but the paracone is reduced in size, less crescentic, its apex relatively more external, and anterointernal to it, below its own anterior crest, the anterior crest of the protocone has unusual strength and distinction. M^1 of the less advanced coryphodonts is intermediate between this and the more aberrant M^{2-3}. The paracone apex is the relatively still more external, but this cusp retains traces of its former crescentic nature; the anterior wing of the protocone crescent is still stronger and tends to sever connection with the paracone, and the posterior wing of the metacone is somewhat weaker. In its next stage, the paracone is styloid.

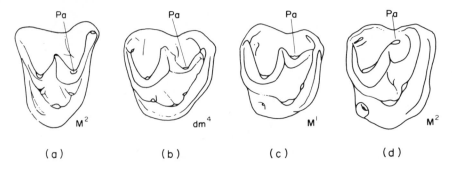

FIG. 2-11. Evidence as to structural evolution and homologies of upper molars in the Pantodonta. A, second right upper molar of *Pantolambda*, a middle Paleocene pantodont with molars little changed from the pattern primitive for placental mammals in general. B, the last (most distal or posterior) right upper milk tooth of *Coryphodon*, an early Eocene pantodont. C, first right upper permanent molar of *Coryphodon*. D, second right upper permanent molar of *Coryphodon*. The series A to D is a structural progression that indicates that the main anteroexternal (or buccal) cusp of the specialized *Coryphodon* molar is surely the homologue of the primitive paracone.

The posterior crest of the *Coryphodon* molar cannot be called the ectoloph, for it represents only the anterior wing of the metacone crescent, and it cannot be said to have rotated, for its relations are exactly as in *Pantolambda*. Nor is the anterior loph a typical protoloph, for it is only the anterior wing of the protocone crescent and in its final development has no connection with the paracone.

The triangular upper premolars of the coryphodonts and those of the uintatheres are superficially similar, but the late Paleocene forms seem to show that the main triangles are not composed of homologous parts in the two groups. In *Coryphodon*, aside from the internal heel, the triangle is composed of the paracone-metacone internally and stylar cusps externally—it is an ectotrigon, homologous with the outer part of the molars. In *Uintatherium* the triangle is composed of the protocone internally and of the paracone and metacone externally—it is an endotrigon, homologous with the inner part of the

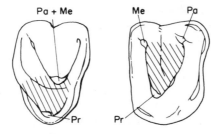

Fig. 2–12. Homologies of right upper premolars of *Coryphodon*, left, and *Uintatherium*, right. The shaded areas are homologous in the two and the two kinds of premolars have evolved in quite different ways.

molars. Here, as in the molars, the resemblance of *Coryphodon*, such as it is, is rather with the most specialized uintatheres than with the primitive *Probathyopsis* or *Prodinoceras*.

The lower molars, while less strikingly different, also have separate histories in the coryphodont and uintathere phyla. In both, the anterior loph is a true metalophid, formed by the protoconid and metaconid, but this is a tendency so nearly universal in ungulates as to be no indication of affinity. The heel is distinctive. In *Coryphodon* the talonid loph runs anteroexternal–posterointernally and is a true hypolophid, joining the hypoconid and entoconid. In *Uintatherium*, on the contrary, the main heel loph runs posteroexternal–anterointernally, and, as shown in *Probathyopsis* where it is incompletely

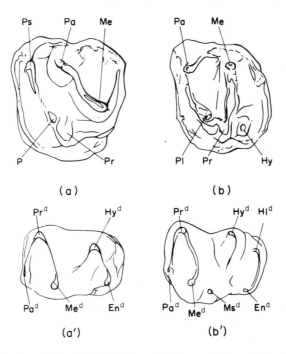

Fig. 2–13. Independent crest (loph and lophid) formation in coryphodonts (Pantodonta) and uintatheres (Dinocerata). A, left upper molar of *Coryphodon*. A¹, left lower molar of *Coryphodon*. B, left upper molar of *Uintatherium*. B¹, left lower molar of *Uintatherium*.

formed, unites the hypoconulid and entoconid. There is also a secondary crest, nearly parallel to this, which runs from the hypoconid toward the metastylid. The latter cusp, absent in *Coryphodon*, is already well developed in *Probathyopsis*.

Both coryphodonts and uintatheres thus represent early attempts at the formation of lophiodont molars, but each has formed these in its own way, converging toward a common adaptive type. The lophiodont perissodactyls represent a third and more successful acquisition of the lophiodont habitus, and they, again, form the lophs in a way peculiar to themselves. The lophs of these three groups are neither homologous nor homoplastic.

The coryphodonts, uintatheres, and perissodactyls seem to me to afford a striking example of the view that when different phyla acquire similar habiti some time after their separation from a common ancestry, there is no inherent tendency for these modifications to arise in the same way in the independent lines. The homologous parts are those alone which were already present in the common ancestry—in ungulate upper molars generally, only the three primary cusps and possibly the conules. There may be a tendency (to call it inherent would involve a personal definition of the word) to form lophs, for instance, but there appears to be no fixed tendency for these to form in the same way or from the same parts in independent groups, except when these phyla had identical or closely similar molars at or after the inception of lophiodonty. Animals which fulfil this last condition are usually closely related, and an apparent inherent tendency to form lophs in the same way in related phyla is consequently often seen, but I conceive the conditioning factor to be not the metaphysical one of germinal predestination but the physical one of mechanical resemblance.

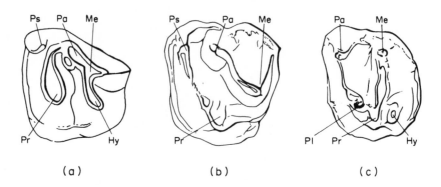

Fɪɢ. 2–14. Independent crest formation in left upper molars of three distinct orders of ungulates. A, Order Perissodactyla, an early rhinoceros. B, Pantodonta, an advanced coryphodont. C, Dinocerata, an advanced uintathere.

If the above conclusions regarding molar evolution in the coryphodonts and uintatheres are correct, it is necessary to suppose that when true lophiodonty was first appearing in the two groups the upper molars of their ancestors, although tritubercular in both groups, were of different mechanical types. For the coryphodonts one would infer an ancestry with strongly selenodont molars, the paracone and metacone apices median. For the uintatheres, projecting backward the *Eobasileus–Probathyopsis* line, one would infer an ancestry with the paracone and metacone noncrescentic, their bases connate, their apices external, the protocone subcrescentic but its crests running to the bases of the paracone and metacone, not anterior and posterior to them.

So far as the dentition is concerned, *Pantolambda* is an almost ideal ancestral type for *Coryphodon*, and it has long been recognized that they are related. Direct ancestry is impossible, for while it would form a suitable basis for the coryphodont dental specialization, *Pantolambda* does not show any actual beginning of this advance, and, judging from the rate of evolution of other phyla, the time between the Torrejon and Clark Fork was much too short to accomplish this degree of evolution.

Source

(1929) A New Paleocene Uintathere and Molar Evolution in one Amblypoda, *Amer. Mus. Novitates*, No. 387, pp. 1–8.

An Exercise in Functional Inference and Model-making

Comment

Devoted paleontologists do not think of fossils as inanimate objects or even as dead. They are remains or traces of live organisms, and they come alive again in the minds of those who study them. That form of resurrection is fairly simple in some cases. If a fossil mammal, for instance, belongs to a living family such as the cat family, Felidae, and resembles some living members of that family in build and dentition, it is a reasonable inference that it also resembled them (within limits not too wide) in behavior and diet. If a fossil belongs to some other family, even one that is extinct, but has cat-like teeth, it is again reasonable (with limits a bit wider) to infer that it was an active predator and lived on mainly vertebrate prey.

If, however, an extinct animal has characters that have no clear and close parallel among those still living, more imaginative methods of investigation come into play and there are likely to be differences of opinion and controversy. That is true, for instance, of the sabertooth cats, the machairodonts (that name comes from Greek roots that mean "sabertooths"), including notably the widespread and popularly known genus *Smilodon*, so abundant in the Rancho La Brea tarpits of Hollywood. Saber-like canine teeth also evolved independently in at least two other extinct groups, but no living mammalian carnivore has canines anything like these. This is an enticing problem to study, and it is here given as an example of such investigation, including a form of model-making. It impressed me when I was making some other studies of early North American cats, and I was further impelled by the then recent 1940 publication of a study on this subject by the Swedish paleontologist Birger Bohlin.

As will be seen, my conclusions were strongly opposed to Bohlin's. Some years later, in 1947, he returned to the fray, but as he seemed neither to meet my objections nor to adduce new evidence, I did not continue the controversy.

My paper is here given in full, with only the slight changes appropriate for this different publication. The approach to the problem and the methods used are here more important than the detailed treatment of a particular example.

Text

The development of saber-like upper canines in flesh-eating mammals has occurred independently at least three times, in the South American Pliocene marsupial *Thylacosmilus* (see Riggs, 1934), in the North American Eocene creodont *Apataelurus* (see Scott, 1938), and in the wide-spread Oligocene to Pleistocene machairodont cats. The machairodonts, at least, survived for millions of years and evidently found this structure eminently useful, but no living carnivore has anything of the sort. The bizarre and savage appearance of these teeth stimulates lay and scientific interest and invites speculation as to their function. The absence of any close recent analogy makes inferences difficult. It is not surprising that a large literature has grown up on the subject and that strong differences of opinion exist.

In 1853 Warren gave an excellent functional description of the canines and associated structures in *Smilodon*. He concluded that the canines transfixed the prey and then cut and tore by movement of the head backwards. He thus foreshadowed both of the views variously emphasized by most later authors: that the canines were used for stabbing (transfixing) and for slicing (cutting and tearing). The stabbing theory was more explicity developed by Brandes whose arguments were, however, weakened by the suggestion of adaptation for preying on glyptodonts. Of course, the fact is that the armor of the glyptodonts and the canines of the machairodonts were fully developed millions of years before these two sorts of animals ever came in contact with each other. Matthew suggested the most widely accepted theory: that the usual prey was the thick-skinned ungulates, the action being to strike and then rip or gash so that the prey bled to death. This view was again and more fully developed in his later paper (Matthew, 1910), which is the classic and still the most useful treatment of this subject. Many later authors, for instance, Scott and Jepsen, have emphasized stabbing or snake-like striking as the only or at least the most important function.

The most recent study is that of Bohlin (1940) who rejects the stabbing theory entirely and maintains that the sabers were ill-adapted for this function and must have served mainly or wholly for slicing. He concludes that the machairodonts were not predaceous but were primarily carrion-feeders, an idea previously suggested by Weber, Marinelli, and a few others, although the great majority of students have believed that all these sabertoothed flesh-eaters were eminently predatory.

Bohlin's study is so able and reasonable, despite its unusual conclusion, that it seemed at first reading to be definitive and to require radical revision of current ideas of sabertooth habits and history. It was, however, then noticed that certain of the arguments advanced by Bohlin as conclusively opposed to the stabbing theory apply with equal force to many of the front-fanged venomous snakes and could be used to "prove" that they cannot strike a stabbing blow or that their fangs are not adapted to this function, which is absurd. Evidently there is a fallacy either in Bohlin's arguments or in the analogy with these snakes, and reconsideration is necessary.

The snake analogy is imperfect and must be carefully limited. The primary purpose of insertion of venomous snake fangs is different from that of the mammalian sabers, which certainly did not inject poison. The associated musculature and osteology are also very different in snakes. Some snakes do, however, insert the fangs by an efficient stabbing action and these fangs are usually, if not always, definitely more curved than they should be if Bohlin's argument is correct, and they are frequently inserted at a "wrong" angle. To this extent the analogy is valid. Relatively little detailed study seems to have been made of the exact motion of snake fangs in penetrating the prey. The best that I have seen (kindly brought to my attention by C.M. Bogert) is by Klauber and refers to rattlesnakes, the fangs of which are less analogous to mammalian saberteeth than are those of some of the other pit-vipers (e.g. *Bothrops*) and many of the elapids. Consideration of all of these reptiles yields a vital clue: the fangs are driven in with a rotary movement and the mechanical center of this rotation does not necessarily or usually coincide with an anatomical joint.

Experiments were made with the skulls (originals or casts) of numerous machairodont cats and of *Thylacosmilus*. (The skull and sabers are unknown in *Apataelurus* although the presence of the latter is certain from the structure of the lower jaw.) These were supplemented by the manipulation of cut-outs and the study of successive tracings made as these were moved in various ways. The general anatomy of *Smilodon* was also considered, along with the special features of musculature distinguishing it from the felines. Some of the results of this study are summarized in the following paragraphs.

Fundamental considerations are the mechanical nature of the canine and the points most conveniently used to represent and analyze motion of the head, in which the canines are immovably fixed. The axis of the canine may be taken as a curved, longitudinal line approximately in the center transversely (labiolingually) and at or near the part of the tooth that is thickest transversely. This thickest part is seldom at the center of the tooth anteroposteriorly but usually more anterior. In *Thylacosmilus* it is marked by a definite ridge, on the labial side, but in the machairodonts the lingual and labial canine surfaces are smoothly curved. In most cases it is noticeable that this axis is more nearly parallel to the anterior than to the posterior margins of the tooth, especially in the more proximal part. The posterior margin is invariably formed by a more acute angle than the anterior and it bears a sharp cutting edge nearly or quite to the alveolus. Although more obtuse, the anterior margin is also always trenchant, at least in its distal portion. A variable extent of the proximal part of this margin is usually rounded and not trenchant.

The points used to analyze head motion are (a) the tip of the canine, which leads the work, (b) the occipital condyles (or the projection of a transverse axis approximately through their centers), which are the mechanical fulcrum nearest to the canines and the point of application of motion from neck and body, (c) the approximate center of gravity, the motion of which is related to the general direction of momentum, and (d) the center of rotation. In distinction from Bohlin's analysis, it is again emphasized that the center of rotation is not necessarily a mechanical fulcrum or anatomical joint. It is an imaginary point helping to visualize and analyze action of a whole series of real fulcrums no one of which is likely to be quite stationary, all the joints of the axial skeleton, especially those of the neck, and also to some extent all the limb joints, especially those of the fore-limb. The center of rotation may be at almost any point within or outside of the animal.

Circumlocution can be avoided by a few simple definitions for present purposes. A pure stabbing motion means a stroke in and then out with a minimum of cutting consistent with the form of the weapon. Cutting is used in the ordinary sense, but for distinction the word slicing is used to mean cutting not necessary merely to insert and withdraw the weapon. Down-slicing is such extra cutting performed during the insertion of the weapon and up-slicing during its withdrawal. Straight-slicing is performed with the weapon remaining at approximately the same depth in the wound.

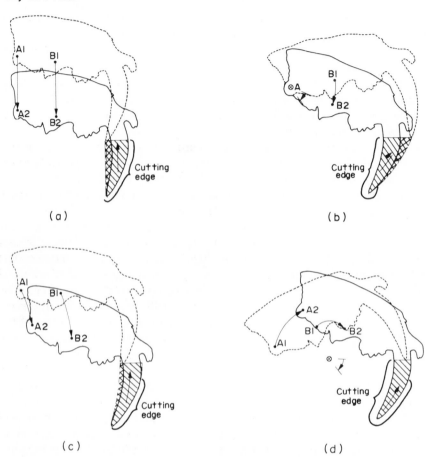

FIG. 2–15. Diagrams of possible methods of stabbing by the sabertooth cat *Smilodon*, discussed in the text. A, motion of head straight down. B, motion by rotation on joint with neck. C, motion by rotation on joint of neck with body. D, motion by rotation around the center of curvature of the canine tooth. The skull is shown in broken outlines at the beginning of insertion of the canines, in solid outline at the completion of insertion. The shaded areas indicate the amount of cutting involved. The arrow shows the direction of maximum strain on the canine. A1 and A2 are positions of the condyles of the skull, its joints with the neck. B1 and B2 are positions of the approximate center of gravity of the head. In D the small circle with an X is the center of curvature of the canine and the arrow near it shows the angular amount of motion involved in stabbing by a motion rotary around that center.

The simplest stabbing motion would be straight down, Fig. 2–15A. This involves considerable slicing, which represents a loss of effort in pure stabbing; it puts great oblique strain on the delicate tip of the tooth; and it uses almost the whole of the anterior edge and none of the posterior edge for cutting. It is obvious that the posterior edge is the more efficient for cutting and that the proximal part of the anterior edge is often incapable of cutting. It seems quite impossible that the animals habitually stabbed in this way (which would, moreover, be a difficult motion to impart to the head with any force).

Rotation about the condyles, Fig 2–15B, also involves some (but less) slicing. The stress is somewhat oblique, although probably not enough to cause unbearable strain. Cutting is mostly on the posterior edge, as it should be, with a little on the distal end of the anterior edge, which is also morphologically justified. Such stabbing was entirely possible, but Bohlin is clearly right in concluding that the anatomical structure is not perfectly adapted to it. Aside from the probably insignificant loss of cutting efficiency, the principal disadvantage is that power is derived only from the muscles depressing the head on the atlantal joint. These are powerful, but the loss of any help or momentum from post-cranial movement would be inefficient. It is incredible that the animals should have tended to hold the atlas quite motionless while moving the head violently.

Bohlin believes that stabbing motion would really be by rotation on the cervico-dorsal joint (Fig. 2–15C), and he concludes that the canines are ill-adapted for stabbing by such a motion. A small amount of slicing is involved. The stress on the tip is oblique, but only slightly and probably not dangerously. The serious disadvantages are that cutting is entirely on the anterior edge of the canine, and that little or no action by the head-depressing muscles is involved, although these are known to have been unusually powerful and can only be inferred to have been hypertrophied by functional relationship to the canines. I emphatically agree with Bohlin that the sabertooths did not habitually stab in this way, while disagreeing just as emphatically with his conclusion that they therefore did not stab at all.

Rotation about the center of a circle of which the axis of the canine is an arc, Fig. 2–15D, imparts a perfectly efficient pure stabbing motion to the canines. There is no slicing and the cutting incidental to pure stabbing uses the whole of the posterior and about two-thirds of the anterior edge of the canine. The proportion of the anterior edge involved in cutting with this motion varies according to the shape of the tooth. In *Thylacosmilus*, for instance, only a small distal part of this edge would need to cut, while in some machairodonts almost the whole edge would be involved. It is not likely to be a coincidence that in each case the part of the anterior edge that must cut for greatest efficiency with this motion does in fact have a cutting edge while the more proximal, theoretically noncutting part does not.

The relation of this motion to the musculature is also advantageous. The motion would result from a thrust of the head forward and its simultaneous depression. Although the occiput tends to rise somewhat, no lifting muscular effort is involved; this is a mechanical result of forward thrust against the occiput accompanied by strong depression of the head on the occipito-cervical joint. A forward lunge is the universal motion of attack, and one of which the sabertooths were certainly well capable. The depression of the head on its fulcrum is, as already noted, provided for by unusually strong muscular development in just these animals. Bohlin adduces as evidence against stabbing that the attacking sabertooth would have to come to a full stop before stabbing and then start a new motion at right angles to the momentum of attack. On the contrary, my analysis seems to show that the attack and the stab would follow naturally and that much of the attacking momentum would be utilized in driving in the sabers. Instead of being less, this seems to me a mode of attack decidedly more efficient than in the "normal" (non-sabertooth) carnivores, in which the forward momentum helps little or not at all to drive in the teeth.

Merriam and Stock state that the canine curvature in *Smilodon* makes it "probable that the downward stroke of the tooth must have been accompanied by a backward jerk in order to make it fully effective." If by "backward jerk" a rotation of the head downward on the occipital condyles is meant, this agrees with the result reached in this paper, but it is to be emphasized that backward motion transmitted to the head from the neck would result mainly in slicing, as discussed later. Merriam and Stock agree that these teeth are chiefly stabbing weapons, and pure stabbing demands that the motion transmitted to the head should not be backward.

This mode of stabbing theoretically requires that the center of the curvature of the canine axis should be well above the canine tip (when the skull is oriented with the palate approximately horizontal), at or a short distance below the alveolar level, below and anterior to the condyles, and posterior to the canines by about half the distance to the occiput or somewhat more. In all the sabertoothed animals that I have investigated, including several genera of machairodontines and the quite independently evolved sabertooth marsupial *Thylacosmilus*, this point is indeed in the region so delimited (Fig. 2–16). Within this region its exact position varies considerably, even by individual

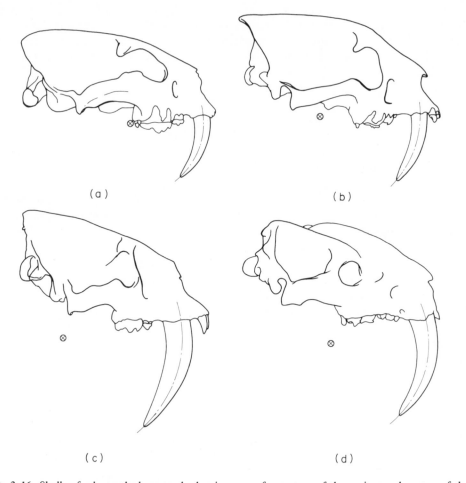

(a)   (b)

(c)   (d)

FIG. 2–16. Skulls of sabertoothed mammals showing axes of curvature of the canines and centers of their curvature. A, B, and C are three different machairodont members of the cat family, Felidae. D, is a South American marsupial, *Thylacosmilus*, that evolved functional machairodonty completely independently of the machairodont cats, a remarkable example of evolutionary convergence.

variation within one species. It is often modified by post-mortem crushing, and it may also be subject to progressive phylogenetic adaptation—a point worthy of more detailed investigation. Canines with the center of curvature well posterior to the head, which Bohlin considers theoretically necessary for efficient stabbing, would give little use to the head-depressers which are apparently the most important head muscles in all these animals. It thus seems that less curved canines would really be much less efficient stabbers than the canines that really occur. It is also significant that venomous snake fangs, which certainly are efficient stabbing weapons and the mechanics of which are analogous to those of saberteeth without being exactly the same, have the center of curvature in a region mechanically analogous to that found in these mammals.

In short, I believe that stabbing in this way is the action for which the whole sabertooth structure is most efficiently adapted. Doubtless the completely perfect stab of the diagram was as rare as perfection always is in nature, but this seems to be the true norm and the adaptive goal of this type of structure.

The more extended final phases of such an idealized normal attack, subject to the myriad deviations of circumstances, are diagrammatically represented in Fig. 2–17. Here motion from *1* to *2* is the final phase of approach in the attack and phase *2–3* the stabbing thrust of the canines. In phase *1–2* the body is moving forward, the neck and head whipping downward and forward. The head is rotating slowly but with increasing acceleration. At *2* the canines hit the prey and a pivoting motion begins—the sudden encountering of resistance on this fulcrum would itself tend to initiate and to assist the pivoting (as would also the contact of the open lower jaw). Forward and downward motion of the head as a whole continues, modified by the still greater acceleration and snapping contraction of the head-depressers as the canines are driven in along their axes. The head might be said to rock forward on the canines, and so transmits in this new direction the greater part of the momentum of the approach.

As far as I know, the only direct evidence of use of a canine saber is that described by Scott and Jepsen, a skull of *Nimravus* with a wound exactly such as would be produced by pure stabbing motion by the associated sabertooth *Eusmilus*. One example does not necessarily indicate a habit, but it does prove a possibility. Incidentally this wound was inflicted on a living animal and not on carrion, a point to be discussed later.

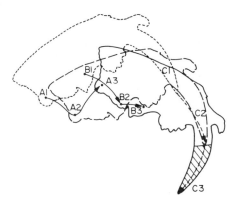

FIG. 2–17. Hypothetical diagram of the skull in final phases of attack by *Smilodon*. Dotted outline, phase one, approach. Broken outline, phase 2, beginning of stab. Solid outline, phase 3, end of stab. A and B as in Fig. 2–15. C, end of canine.

Consideration should also be given to the possibility of purposeful slicing, as opposed to slicing incidental to deviations from perfect stabbing. Most adherents of the stabbing theory have considered slicing as a probable accompaniment or secondary function. Bohlin considers it the primary function. In the first place, slicing without stabbing seems unlikely. The canines cannot slice unless they are inserted in the hide or flesh, and the easy way to get them in would be by stabbing, whether in its pure attack form or in some modified manner.

Straight slicing considered as an activity independent of stabbing is diagrammatically shown in Fig. 2–18A. Personal experience of the practical impossibility of cutting tender meat with a sharp knife without hacking or sawing motion is suggestive of the tremendous effort necessary for a sabertooth to slice tough hide and muscle in this way with its duller weapon. In view of the very oblique strain involved, it is doubtful whether the teeth could stand continual application of this great force even if the animal could exert it.

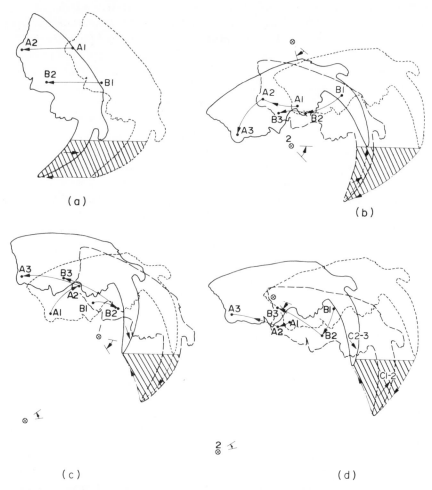

(a)

(b)

(c)

(d)

Fig. 2–18. Diagrams of slicing by *Smilodon*. A, straight slicing. B, down-slicing. C, up-slicing. D, down-and-up slicing. Symbolism and construction as in Fig. 2–15. Arrow on canine indicates direction of greatest stress in the phase immediately before the position of the canine shown.

It seems probable that a sabertooth with the desire or necessity of slicing with the sabers would adopt the expedient of hacking, that is, of combining stabbing and slicing (a straight chopping motion against the edge being out of the question). Such stab-slicing could involve down-slicing (Fig. 2–18B), up-slicing (Fig. 2–18C), or both (Fig. 2–18D). In all three figures the amount of slicing is roughly the same. In down-slicing the head is jerked backward while being rotated (or depressed) about a center near or above the top of the posterior part of the skull. The saber is then withdrawn without cutting by rotation in the opposite direction around the center of curvature of the canine. The oblique strain, especially on the tip of the canine near the beginning of the cut, would be great. If a slice of any considerable extent were made at one stroke the direction of motion would tend to be nearly at right angles to the cutting edge, minimizing shearing or sawing action and making the cutting very difficult.

In up-slicing the canine is first stabbed in by pure stabbing without slicing and is then withdrawn by pulling the head backward and upward, or with rotation about a point below and behind the center of canine curvature. (In such a case a diagram showing rotation on a fixed point is doubtless over-simplified; the center of rotation probably moved and the true course might be more nearly a section of a spiral than an arc of a circle.) This would probably require somewhat less effort than down-slicing, because the difficult insertion is accomplished in the easiest possible way and the cutting could largely be performed by rocking backward on the fore-limbs against the gradually decreasing pull of the head-depressers. The effort would nevertheless be great and a long slice would require almost prohibitive exertion and transverse strain on the tip of the canine.

A down-and-up slice can be accomplished by relatively straight backward motion at the condyle, accompanied by depression and raising of the head, mechanically equivalent approximately to rotation first about a point near the posterodorsal end of the skull and then rotation in an opposite direction about a point below the posterior end of the skull. This second rotation would follow almost automatically from the backward pull against the cutting resistance. Such motion is smoother and easier than either pure down-or-up-slicing, and it seems psychologically and physically a more natural motion for an animal to make. For a given size of slice, its efficiency would also be greater because the angle of the cutting edge to its work would be more acute, increasing the relatively easy sawing or shearing component, and its safety would be greater because less (although still considerable) strain would be put on the tip of the tooth. On the whole it seems probable that if the animal did intentionally slice, it would tend to do so in this way.

The result of such an action is somewhat like that of a stab that is unnecessarily broad. As typically developed, however, the down-and-up slice is unlike the pure stab. In the former the neck or body motion is backward, in the latter predominantly forward. One would be used more on disabled or dead prey, the other more in attack on active prey. A sabertooth in action would not be concerned with following diagrams or maintaining maximum efficiency, and undoubtedly would use all sorts of variant and intermediate motions, but pure stabbing and down-and-up slicing seem to be the norms of the two most likely uses of the canines.

Another factor to be considered is the possible effect of preponderance of strain at the alveolus in any one direction other than along the tooth axis. It is well known that such pressure has a tendency to make the alveolus move. Thus under normal conditions the alveolar, and hence also the coronal, orientation may often be inferred to be one such that the tooth axis represents the average direction of pressure. In any sort of stabbing except that shown in Fig. 2–15D, the average pressure is not along the tooth axis. Pure stabbing as in Fig. 2–15D is the only likely action that centers strain in the direction of that axis, and therefore is probably the normal use correlated with the observed orientation. Any probable sort of slicing is almost certain to produce a pressure component tending to rotate the crown forward about the alveolar mouth as a fulcrum or to move the alveolus itself forward. For a given amount of slashing, this component would probably be weakest or least constant in down-and-up slicing as in Fig. 2–18D. It is improbable that the observed orientation was developed in relation to slicing of any sort as the predominant use of the teeth, although probable that a moderate amount of slicing as a secondary activity would not seriously disturb the orientation correlated with stabbing. Other factors enter into the situation (such as a growth or other tissue pressure counteracting the pressure of use) and some animals do habitually use teeth with pressure oblique to their axis and predominantly in one direction. The argument cannot, in itself, be considered conclusive, but it is at least suggestive.

Source

(1941) The Function of Saber-like Canines in Carnivorous Mammals, *Amer. Mus. Novitates*, No. 1130, pp. 1-12.

Other references

BOHLIN, B. (1940) Food Habit of the Machaerodonts, with Special Regard to *Smilodon, Bull. Geol. Inst. Upsala*, Vol. 28, pp. 156–174.
BOHLIN, B. (1947) The Sabre-toothed Tigers Once More, *Bull. Geol.; Inst. Upsala*, Vol. 32, pp. 12–20.

How a Strange and Ancient Animal Got About

Comment

Another approach to the study of fossils as living animals is to reconstruct parts of the musculature and to study their mechanical functions. Although this is possible and interesting in some invertebrates such as the bivalves, it has been more extensively applied to vertebrates. In them most of the actions involved in behavior are made by the pull of muscles that are attached to the skeletal parts at both ends. Traces of the muscles themselves are very seldom preserved, but it is often possible to determine the places and areas of their attachments on the bones, or in fewer instances cartilages, which are preserved and to which the muscles were attached.

Such reconstructions have been made for some members of all the vertebrate classes. The example here given is for a mammal of the extinct group Multituberculata, which endured for some 175 million years (Jurassic to Eocene) but then became extinct. In characters of the dentition, skull, and brain (mentioned and figured earlier in this chapter), these mammals were quite distinct from any others and inferences as to their ways of life present a severe challenge. Their postcranial characteristics are still incompletely known. The study from which this extract is taken is based on a unique specimen comprising the pelvis and hind limbs of the genus *Eucosmodon* found in earliest Paleocene rocks, about 65 million years old, in the San Juan Basin of New Mexico.

This study was made in collaboration with Herbert Elftman, a comparative anatomist then a graduate student and later on a member of the medical faculty of Columbia University. Elftman had made many dissections of the limb musculature of a wide variety of living mammals and brought that expertise to the study of one long dead. For present purposes our identifications of individual muscles do not particularly matter. The point here is the method used and the inferences drawn. In the paper as a whole other knowledge of multituberculates was also discussed in making broader inferences as to the habits of this large, strange group as a whole and of *Eucosmodon* in particular. Here part of the section on limb musculature and function is reproduced without change and the final summary is also excerpted and included here. No more recent study of this subject has appeared.

Text

In restoring the musculature, accurate restored models, three times natural linear dimensions, were made of the whole pelvis and of the right femur, tibia and fibula. These were mounted in a natural standing pose and the muscles were then fashioned in red modeling clay and affixed in the inferred original positions. Constant reference was made to the original bones, on which the chief muscle attachments are clearly visible, to numerous dissections of recent mammals by Elftman, to the large literature on recent myology, especially of the marsupials and monotremes, and also to the small but important literature on paleomyology (especially Gregory, Camp, Romer, etc.). The completed restoration was then studied, drawn, and dissected as if it had been a recent mammal.

The mechanical effect of the limb musculature depends in large part on the normal posture of the various bony elements. This posture is to be inferred chiefly from the osteology itself, the relative sizes of the different segments, the morphology of the individual bones and, especially, the shape and extent of each articular surface. As all of these surfaces are preserved in the specimen here especially studied, the limb posture is known with a reasonable degree of certainty and serves as a point of departure for a functional analysis of the musculature. The femur had unusual freedom of motion in all directions, but in the normal standing position of the animal it was nearly horizontal, inclined forward, outward, and slightly downward. The angle between the femur and tibia could never have exceeded 90° in life and was usually considerably less—the leg could not be straightened and the crus was usually drawn well backward and could be made almost parallel with the upper limb. Normally the tibia and fibula would thus be directed backward, inward, and downward. The foot, as clearly shown in its almost completely known osteology, is unusually primitive. It is pentadactyl and plantigrade, of grasping type with partially opposable hallux. The digits are strong, the functional length formula $3 > 4 = 2 > 5 = 1$. The terminal phalanges carried claws of moderate size, somewhat compressed transversely.

The foot was clearly a rather flexible structure capable of much movement in all usual

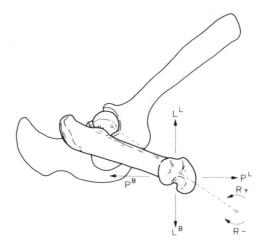

Fig. 2–19. Reconstruction of pelvis and right femur of *Eucosmodon* showing a normal position and the components of possible motions of the femur. Arrows show directions and symbols results of motions: LB, levation of body; LL, levation of limb; PB, protraction of body; PL, protraction of limb; R+, positive rotation; R−, negative rotation.

ways. The motion of the crus, aside from possible relatively slight rotation, was chiefly simple flexion and extension. The most complex and important movements are those of the femur. These movements are of two quite distinct sorts: rotation of the femur in all directions about a fixed point, the head, and rotation about a linear axis—a line through the center of the head and the middle of the distal end (see Fig. 2–19). Further analysis shows that all possible movements of the femur may be resolved into six components, as follows:

(A) Rotation about head of femur:
(a) Horizontal components (protraction):
    (1)   Forward—protraction of limb.
    (2)   Backward—protraction of body.
(b) Vertical components (levation):
    (3)   Upward—levation of limb.
    (4)   Downward—levation of body.
(B) Rotation about linear axis:
    (5)   Counterclockwise as viewed from distal end of right femur—positive rotation.
    (6)   Clockwise as viewed from distal end of right femur—negative rotation.

In considering a single limb during ordinary straightforward locomotion two effects alternate: (1) the distal end of the femur moves forward relative to the pelvis, and (2) the pelvis then moves forward relative to the distal end of the femur. It is clear that these effects are purely relative and that the resultant motion, involving as it does some lateral motion of the pelvis relative to the substratum, is actually rather more complex. The pelvis, and hence the body, is furthermore moving forward at a nearly constant rate, as protraction of the limb against the body on one side is simultaneous with protraction of the body against the other limb. Each of these two phases of the motion of a single limb involves three of the six components of femoral movement:

First Phase	Second Phase
Protraction of limb	Protraction of body
Levation of limb	Levation of body
Positive rotation	Negative rotation

As will appear in the subsequent functional classification of the individual muscles, the majority of the muscles are involved in more than one of the six components. (Reconstruction of the muscles is shown in Figs. 2–20, 21, and 22). In most cases, however, a given muscle is involved in only one of the two groups above, that is, in only one of the two phases of the normal stride. Exceptions are seen in the femoro-coccygeus, which is a protractor of the body but also a levator of the limb, and in the obturator externus, which is a protractor and levator of the body but also a positive rotator.

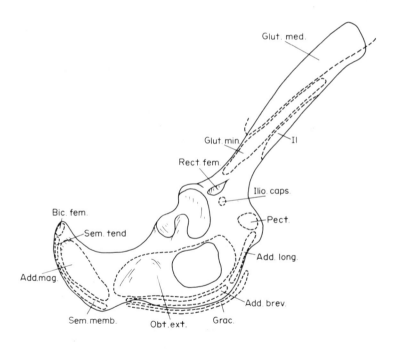

FIG. 2–20. Right lateral view of restored pelvis of *Eucosmodon* with inferred areas of muscle attachment. Anatomists will readily recognize the abbreviations of names of muscles. The general purpose of the figure here is to show a step in the reconstruction of musculature in an extinct group of animals.

The terms "protractors of limb or body" and "levators of limb or body" seem to us preferable for use in functional analysis to the anatomical terms "flexors," "extensors," "adductors" and "abductors" which consider the body as if it were suspended in space (or laid out on the dissecting table.)

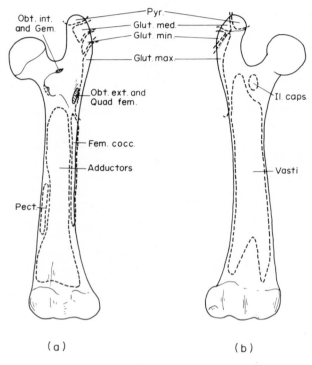

FIG. 2–21. Views (A, ventral; B, dorsal) of the right femur of *Eucosmodon* with inferred areas of muscle attachment. Purpose as for Fig. 2–20.

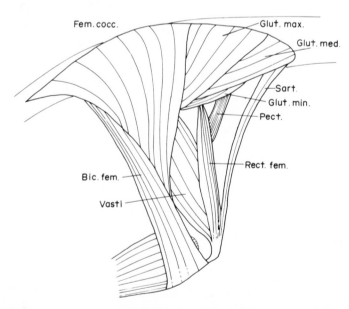

FIG. 2–22. Sketch of the restoration in modeling clay of the superficial muscles of the right hind limb of *Eucosmodon*. The deeper musculature was also restored and sketched by dissecting the model.

Rotation of the femur about a linear axis is unusually important in *Eucosmodon*. The muscles bringing about this rotation are powerful and their origins, courses, and insertions are such as to make their action more effective than it is in most mammals. As brought out below, the lesser trochanter serves almost wholly for the insertion of positive rotators and is strongly developed and specialized in such a way as to give strong leverage for this motion. The greater trochanter, chiefly for insertion of the glutei, is analogously developed for negative rotation. It results from the nearly horizontal femur and the acute angle between thigh and crus that positive rotation of the femur, without other simultaneous movements, would cause movement of the foot forward and somewhat inward. Negative rotation, with the foot planted on the ground, would result in moving the body forward and slightly away from this foot.

The position of the femur together with this strong adaptation for rotation distinguishes *Eucosmodon* rather sharply from most other mammals and from most reptiles. In the majority of mammals the distal end of the femur was drawn more anteromedially beneath the body and rotation, while often present, is generally much more limited than in *Eucosmodon*. An essential difference between *Eucosmodon* and most other mammals is that the muscles which are primarily protractors in the former become involved also in levation in the latter, due to the changed position of the femur. The disadvantages of retaining a somewhat more reptilian posture were compensated in *Eucosmodon* to a considerable degree by the freedom of rotation.

The actual movements of the animal, whether in the simplest case of normal straightforward locomotion or in its more irregular activities, such as turning or moving backward or laterally, would obviously each involve a complex combination of the muscular activities here analyzed into their simplest components. The limb was no doubt occasionally used also for purposes other than locomotion (scratching, seizing objects, etc.) but the chief movements involved would be the same except for being related to the pelvis as a fixed point.

Résumé:

(1) The femur of *Eucosmodon* was held in a nearly horizontal position, pointing outward and slightly forward and downward.

(2) The restoration of the pelvic musculature shows it to have a degree of differentiation similar to that of other mammals. [Note added in 1979: the differentiation was found to be quite different from and more advanced than that of the egg-laying mammalian monotremes and although distinctive more like that of marsupial and placental mammals.]

(3) The details of muscular morphology and function are characteristically different from those of other mammals or of reptiles.

(4) The pelvic musculature, in agreement with all other known anatomical and environmental features, indicates an arboreal mode of life.

Source

SIMPSON, G.G., and H.O. ELFTMAN (1928) Hind Limb Musculature and Habits of a Paleocene multituberculate, *Amer. Mus. Novitates*, No. 333, pp.1–19.

3

Paleoecology and Faunal Analysis

BIOLOGISTS who study whole organisms and faunas and floras have always realized that any understanding of them must take into account their relationships with each other and with their physical environments. From Darwin onward evolutionary biologists have been particularly aware that evolution cannot be understood without attention to those relationships, especially as one of the major requirements of evolutionary theory is to recognize, define, and explain the origin and maintenance of adaptations of organisms to ways of life within environments occupied or accessible. The study of ecology has thus become an essential part of the history of life. As fossils and their occurrences are the primary data for study of that history, these also have a historical ecological, for them paleoecological, significance.

When Thomas Jefferson in 1797 (he was not yet president) described some fossil bones as those of a gigantic lion, a fierce carnivore supposed to be still living in the frontier wilds he was making an ecological deduction. That is true in spite of the fact that Jefferson was wrong on all counts. His friend Caspar Wistar soon set him right: the creature was a ground sloth; it was a herbivore; and it was extinct. The incident is trivial in a way, but it is a historical precedent for continuous and still increasing attention to ecological studies by American and other paleontologists. This field might be said to have come of age with the publication in 1957 of an immense (1077 page) volume on marine paleoecology. That and a few other general works are cited below.

It has become almost routine to include some paleoecological considerations in studies of fossil faunas. Descriptions of taxa (classified and named groups of organisms) also often but less routinely include aspects of function and evolution that have a bearing on ecology. The study of footprints and trackways preserved in rocks has become a subdivision of paleontology with a name of its own: ichnology, "ichnos" being the classical Greek word for "footprint" or "track." These have a special status that has a bearing on ecology. Such fossils are not the dead remains of organisms but are evidence of their activities while they were alive. A broader term, "trace fossils," includes burrows, which are surprisingly varied and frequent in the fossil record. Coprolites, the fossilized remains of excrement, are also diverse and common and may give some evidence on food habits.

Several other texts of ecological interest are included in other sections of this book. In this brief section I include just three excerpts of different approaches explicitly paleoecological.

References

Because I have no example to give but consider the subject interesting I have included several references for the somewhat marginal subject of ichnology.

AGEE, D.V. (1963) Paleoecology, McGraw-Hill, New York, San Francisco, Toronto, London.

CASAMIQUELA, R.M. (1964) Estudios Icnológicos, printed in Buenos Aires and distributed by the University and Museum of La Plata, Argentina. (I have included references to this and to Seilacher because I know of no comparably extensive and general discussions of ichnology in English. Agee, cited above, does have one brief chapter on trace fossils of all sorts. There have been innumerable descriptive studies of particular trace fossils, many of them up to about 1960 cited by Casamiquela.)

HÄNTSCHELL, W. (1962) Trace fossils and Problematica, in *Treatise on Invertebrate Paleontology*, (R.C. MOORE, editor) Geological Society of America and University of Kansas Press, Pages W177–W245.

IMBRIE, J., and N. NEWELL, editors (1964) *Approaches to Paleoecology*, John Wiley and Sons, New York.

LADD, H.S., editor (1957) *Treatise on Marine Ecology and Paleoecology*, Volume 2, *Paleoecology*, Geological Society of America, Memoir 67.

LAPORTE, LEO (1968) *Ancient Environments*, Prentice-Hall, Englewood Cliffs, New Jersey.

MURIE, O. (1954) *A Field Guide to Animal Tracks*, Houghton Mifflin, Boston. (Recent, only, and almost all on North American living mammals, but invaluable information on mammal tracks and scats [unfossilized excrement]. Moreover, unlike most field guides, it is not only for those who merely want to list the names of animals. It also includes delightful personal discussion of the lives of the animals.)

SEILACHER, A. (1953) Studien zur Palichnologie, *Neues Jahrb. Geol. Paläont., Abhandl.*, Vol. 96, pp. 421–452, and Vol. 98, pp. 87–124.

VALENTINE, J.W. (1973) *Evolutionary Paleoecology of the Marine Biosphere*, Prentice-Hall, Englewood Cliffs, New Jersey.

Ecological Analysis of an Ancient Local Fauna

Comment

In 1879 a party working for Yale Professor O. C. Marsh under the field direction of William Reed opened Quarry 9 in the Morrison Formation at Como Bluff, Wyoming. That formation was and still is famous for its richness in dinosaurs of late Jurassic Age, about 135 million years ago. Some dinosaurs turned up as bits in Quarry 9, but the great feature of that quarry was a windfall: the largest collection of mammals of the Age of Reptiles that had then ever been made and still one of the richest. Although Marsh published preliminary notes on many of those rare finds, their full and detailed study still had not been done until I undertook it as a graduate student at Yale in the 1920s. I then found that the collection from Quarry 9 was even more extraordinary than I had realized. It included not only the diverse mammals but also possible representatives of every order of nonmarine vertebrates known to have been in existence in the late Jurassic. Thus study of the Mesozoic mammals from there could be enriched by placing them ecologically in virtually the whole community in which they lived.

For present purposes the more detailed faunal list with which I worked can be summarized as follows:

Invertebrates:
    Some snails and freshwater mussels.
Fishes:
    A lungfish and several others.

Amphibians:
A primitive frog.
Reptiles:
Turtles.
A small lizardlike animal, a rhynchocephalian like the tuatera now surviving only on islands near New Zealand.
A crocodilian.
Bits and pieces of at least eight different dinosaurs, some, especially those called laosaurs, herbivorous, and others, among the theropods, carnivorous.
A pterodactyl.
Some doubtfully identified lizards.
A probable but not quite surely identified early bird.
And many small mammals referable to four different orders all now extinct as such although the Pantotheria have living descendants, including us: Multituberculata, Triconodonta, Symmetrodonta, and Pantotheria.

In discussing this fauna as a community I adopted from the German biologist Möbius his term "Biocönose," defined as a "Lebensgemeinschaft" ("life association"), but for use in English I suggested turning the Greek roots around and calling this a "cenobiota," that is, a biota of species that lived in common (Greek *koinos*, which is best Anglicized in the form *ceno*). I still think that would have been preferable English usage, but later ecologists who adopted the concept have usually called it "biocenosis."

A fauna similar in character and of about the same age had been recovered from the "Isle" (actually a peninsula) of Purbeck in southern England, and I made some comparisons with it.

In ecological consideration of this community I concentrated on the energy flow and the relationships variously termed trophic levels, food chains, food webs, etc. When I wrote this paper that concept and diagrams illustrating it were already well known but only in application to living animals. Quite likely someone had also undertaken this for a fossil association, but I do not know of such an earlier publication and in any event this sort of study of fossils was unusual at that time. Since then considerably more sophisticated studies like this have been made.

A point that did not occur to me at the time but that I now think rather likely is that the specimens of mammals, all disarticulated pieces, might well have been what was left of animals eaten by the crocodiles, turtles, and perhaps also fishes. They environment of actual deposition was probably acquatic, and the local concentration of bits and pieces of land animals could have resulted from predation by the aquatic carnivores.

The text has been excerpted with slight changes from a paper considerably longer.

Text

As is often, indeed usually, the case, a simple perusal of the faunal list of Quarry 9 does not given an adequate idea of the actual animal association. Thus, although several large forms, even including *Morosaurus* [one of the big sauropod dinosaurs], appear, they certainly do not belong to this ecological group, while remains of some of the smaller ones are very rare. It is also certain that a number of types of organisms not mentioned in the faunal list were present and they must be included in a study of the fauna even though they are not preserved as fossils.

There are two more or less distinct subfaunal units or cenobiotas, affecting each other very frequently, it is true, but on the whole independent. The first comprise the aquatic and thoroughly amphibious animals, the *Unios* [freshwater mussels], the fish, the frogs, the turtles, and the crocodiles. The other group, the small fauna of the land of the Morrison region, consists of the rhynchocephalians, the small dinosaurs, the pterodactyls, the lacertilians, the birds, and the mammals. All the animals mentioned were probably common enough to be of importance at the time, as those which are very rare in our collections (frogs, pterodactyl, bird) are just those of which the smallest percentage of individuals are usually fossilized.

The first step in the energy cycle which this fauna presents is obviously now missing as it must have consisted of green plants, as in all cases. There are numerous brown to black carbonized or sometimes rusty films and irregular impressions in the clay matrix which must betoken a considerable development of aquatic, and no doubt also terrestrial, vegetation. In the aquatic subfauna there are no exclusive herbivores, so that it is plain that here at least one other step is missing. The mollusks, chiefly *Unios*, fed on microscopic organisms, both plant and animal. The lung fishes fed on small mollusks, on crustaceans, worms, etc. The other fishes apparently did not eat mollusks or the harder and larger crustacea, but must have thrived on small crustaceans, aquatic worms and larvae, and also on such insects as became accessible to them. This complex of small invertebrates has left no fossilized trace, but its abundant presence and general character is as manifest as if all the details were known. The frogs also belonged to the group of animals which subsisted chiefly on this small invertebrate diet. It is probable that the frogs, especially when young, also served in another role less pleasant to them, namely that of a supplementary article of diet for the fishes and small turtles.

Our attention is thus turned to the next ecological element of the aquatic cenobiota, that of the highest carnivorous or vertebrate-eating animals, the ones farthest removed from the beginning of the energy cycle. The fish belong here only to a slight extent, although they undoubtedly ate tadpoles, frogs and smaller fish on occasion. The aquatic carnivores *par excellence*, however, were the crocodiles and turtles. As their limb structure clearly shows, the members of both of these groups were still well able to progress on land, but their chief source of food and their most congenial home was in the water and hence they are properly considered with the aquatic cenobiota.

The rather small crocodiles were surely exclusively carnivorous, eating on the one hand fish, frogs, and other adequately large aquatic animals, and on the other such mammals and small dinosaurs as were so unlucky as to fall into the water or so unguarded as to let themselves be carried thither. The possibility of a connection between the development of the crocodiles and of their prey was first pointed out by Owen in 1879. The earlier true crocodiles were almost exclusively of longirostrine [long-snouted] type, and not until the upper Jurassic did brevirostrine [short-snouted] crocodiles become common. It is a notable fact, also, that in the upper Jurassic the latter were markedly smaller than their contemporaneous long-snouted cousins and also smaller than their Tertiary and Recent short-snouted relatives. That the spread of brevirostrines at this particular time and their small size were associated with a simultaneous expansion of small animals suited to be their prey is a natural assumption. If we compare the food habits of modern forms, it is found that the longirostrines are largely or exclusive piscivorous while the brevirostrines are also in part fish-eating but often depend largely upon higher prey, especially mammals. The conclusion is very tempting that the spread of Mesozoic mammals in the late Jurassic was a factor in the simultaneous spread of small brevirostrine crocodiles. Whether the causal connection was actually so close or not, and one prefers not to commit oneself on such scant evidence, it seems certain that the crocodiles of the upper Jurassic were admirably suited to prey on the small contemporaneous mammals and they quite surely did so. Owen has pointed out that the dwarf crocodiles *Nannosuchus* and *Theriosuchus* are found in actual association with abundant mammalian remains in the Purbeck. The same is true of *Goniopholis* in both the Purbeck and Morrison. *Goniopholis* was rather larger, it is true, and an exceptionally favored individual might reach a length of some nine feet, but the average was considerably less. In the Quarry 9 material most of the crocodiles are of small size, ranging from infants of a foot or so up to adults of five or six feet in length. The largest would have no reason to scorn a mammal or a small land reptile as an article of diet. The skull proportions and dental characters are not significantly different from those of the most inveterate mammal-eaters among recent crocodiles.

As is well known, the turtles of the present day (in a general sense, including also tortoises and terrapins) have the most various food habits. Those which inhabit fresh water, however, and which otherwise seem most nearly comparable to the Morrison forms (*Probaëna* and *Glyptops*) are almost exclusively carnivorous. Despite their inferior size, the food of the Quarry 9 turtles was probably comparable with that of the crocodiles to a large extent, although perhaps including smaller animals and even some vegetation. The ferocity and predaceous ability of the modern snapping turtle is well known.

In the aquatic cenobiota the crocodiles and turtles represent the highest point in the energy cycle. It is very unlikely that they were the prey of any other predaceous carnivores. Not only were they the largest forms of their cenobiota, but both were enclosed in a heavy bony armor, *Goniopholis* not being inferior to any turtle in this respect. After the death of one of these forms the energy stored in his tissues at that particular time would usually be returned at least in part to one of the earlier stages of the cycle.

Turning to the land, we find that the fundamental forms, the land plants, are again unknown from this quarry and very poorly known from any part of the Morrison Formation, but their abundance may safely be assumed. On them fed not only certain of the vertebrates, the laosaurs and the herbivorous mammals (multituberculates), but also, with the intervention of intermediate steps in many cases, all of the land invertebrates. On these latter in turn depended the pantotheres among the mammals and the rhynchocephalians and lacertilians among the reptiles, all of which were apparently of the dietetic type commonly, but rather inaccurately, styled insectivorous. While some may have been strictly insectivorous, the majority must have added to the latter other land arthropods, worms, etc. That no such invertebrate animals are known as fossils from the Morrison is a gap in our knowledge, but does not involve the slightest doubt as to their abundant presence at the time. Insects are a rather common element in the English Purbeckian, a fauna essentially contemporaneous with that of Quarry 9 and of somewhat similar facies.

The smaller theropod dinosaurs and the symmetrodont and triconodont mammals were the predaceous carnivores of the terrestrial cenobiota, and they found their prey in the laosaurs, multituberculates, pantotheres, rhynchocephalians, and lacertilians.

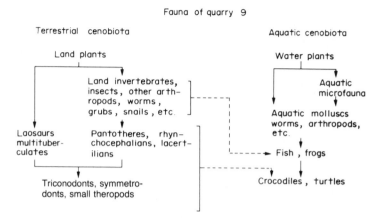

Fauna of quarry 9

Terrestrial cenobiota

Aquatic cenobiota

Fɪɢ. 3–1. Some of the trophic relationships or food chains of the late Jurassic terrestrial and aquatic communities represented by or inferred from the fossils found in Quarry 9 at Como Bluff, Wyoming. The arrows indicate the direction of passage of food and energy among the various members of the local faunas and floras.

There must have been numerous interactions between the terrestrial cenobiota and the aquatic one. Probably the most constant and important of these were in the first place the eating by the fish and frogs of the terrestrial invertebrates, and in the second place (and probably more significantly) the preying of the more properly aquatic carnivores upon the small land vertebrates of all groups, especially the mammals. All the relationships may be summed up in the diagram, Fig. 3–1, it being understood, of course, that such a schematic representation involves the omission of many intermediate steps and of many details and variants in order to bring out the more fundamental features.

The arrows indicate the more accentuated lines of food derivation and energy flow.

Source

(1926) The Fauna of Quarry Nine, *Amer. Jour. Sci.*, Series 5, Vol. 12. pp. 1–11.

Faunal Analysis, Facies, and an Evolutionary Principle

Comment

The following text is part of the discussion in a monographic systematic description of mammalian faunas from the middle to late Paleocene (some 61 to 59 million years before the present) of Montana. This general discussion is largely at the level of orders and families but one of its topics is at the specific level.

Several different subjects are treated in this excerpt, all by methods appropriate for any faunal analyses, including those of living faunas, and all now in one form or another widely used.

The first topic is the makeup of the collection in terms of the relative numbers of individuals in various taxa. The specimens were divided into four lots, by origin, the largest lot from what was designated as the Gidley Quarry, a middle Paleocene assemblage all of exactly the same geological age. The Silberling Quarry yielded mammals of approximately the same age as those of the Gidley Quarry but from about 4½ miles distant and with somewhat different proportions of orders and families, probably a sampling effect but perhaps also involving some difference of facies. Middle Paleocene mammals from other localities in this general area were not from quarries but were found near or at weathering surfaces. As is stated in the following excerpt, they proved to be different in facies from the quarry faunas. A third quarry, the Scarritt Quarry, was higher in the succession of sediments, therefore later in age, and turned out to be markedly distinct, mostly at the generic and specific level, and to belong in the late Paleocene.

The basic faunal analysis in this general approach was by percentages in each of the four lots of specimens belonging to the various families recognized. The numerical data are represented visually in Fig. 3–2 as a pie diagram for the Gidley Quarry alone and in histograms for all four lots in Fig. 3–3, which facilitates comparisons. Later work on Paleocene mammals would somewhat modify the classification used, most noticeably by transferring the family Arctocyonidae from the Carnivora to the Condylarthra, but the interest here is more in the exemplification of the methods than in the details of the examples themselves. This particular approach to faunal analysis in paleontology was later improved by counting not the total number of specimens referred to a given taxon but the lowest number of individuals that could be represented by those specimens. For example, two isolated right femora of a given species count as two individuals but one right and one left femur count as only one individual. This problem does not arise in the usual counts of Recent faunas.

The next point briefly made is that the Gidley and Silberling quarry collections are ecologically similar but the collection from surface localities is ecologically different from them. A third topic touches on the probable behavior and ecology of the major groups present.

The fourth subject involves the accumulation and preservation of the mammal remains in the Gidley Quarry. A point here is that while this collection samples essentially a single biocenosis, that is, a local community of associated and interacting mammals, it is in fact a thanatocenosis, that is, an association of the remains of animals after their deaths. (Greek *thanatos*, "death".) An inference is made as to how this particular thanatocenosis was formed. This is a fairly early example of one aspect of a subject that has become a rather complex paleontological specialty or subscience under the name of taphonomy (Greek *taphos*, "burial," and *nomos*, "law") now broadly defined as the study of everything that has happened to a fossil from the time of death to the time when a specimen is in a collector's hands.

The final part of this excerpt considers the fauna from the Gidley Quarry in the light of what is here called "ecological incompatibility." The general principle involved (I would not now call it a "law") was already mentioned by Darwin in other words, and various aspects have been "discovered" by others repeatedly ever since. As far as I know Matthew, here cited and quoted, was the first to apply it to the taxonomic interpretation of fossils.

Text

The percentage composition of the principal surface localities, all near the same level and similar in facies, and of each of the three quarries is shown graphically in Figs. 3–2 and 3–3.

The figures represent the composition of the identified collections. They are somewhat biased in favor of the multituberculates, since practically all the specimens of these are identified, whereas there are numerous isolated teeth of other orders that are not identified, but in general they are representative of the faunas as they occur.

The Gidley and Silberling Quarries agree well, within the probable limits of chance, except for the much smaller proportion of primates and greater of multituberculates in the Silberling Quarry, a difference probably representing slight local distinction in facies.

The surface localities differ remarkably from the quarries and suggest very different environmental conditions. This result is doubtless somewhat biased by the fact that minute forms, like insectivores and primates, are more likely to break up and are harder to find at surface localities, but this does not explain the difference. Some of the so-called surface material was, in fact, found in place. These localities have been very closely examined and fragments as small as the smallest isolated insectivore or primate teeth recovered, so that the almost total absence of those groups must really result from their great rarity here. Furthermore, the arctocyonids are really much more abundant at these localities, for not only the relative but also the absolute number of specimens is greater for these localities than for the Gidley and Silberling Quarries, despite the much larger collections from the latter. This "surface" fauna, 90 percent carnivores and ungulates, is of more normal type, in comparison with Tertiary faunas generally, than

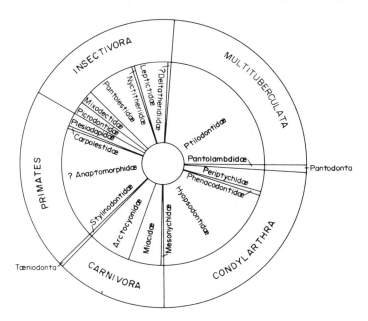

Fig. 3–2. Pie diagram of the relative abundances of identified specimens of mammalian taxa in the National Museum collection from the middle Paleocene of the Gidley Quarry, Crazy Mountain Field, Montana. The outer circle represents the orders as originally classified, the inner circle the families.

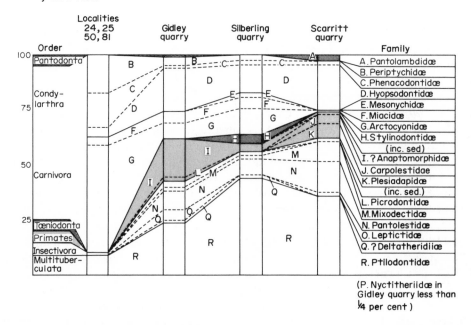

FIG. 3–3. Histograms of individual abundances of identified specimens belonging to orders and families in middle Paleocene surface localities and two middle Paleocene quarries (Gidley and Silberling) and one late Paleocene quarry (Scarritt) in the Crazy Mountain Field, Montana.

are the quarry faunas. Its members average larger than do those of the quarry faunas, and they are probably terrestrial for the most part. This appears to be a normal floodplain facies, rather closely analogous to that of the Torrejon. Its most marked peculiarity is the higher percentage of carnivores than of herbivores, a condition for which no probable explanation is seen.

Even at the surface localities there is a surprisingly low percentage of animals really of large size for the Paleocene. The phenacodonts are of average size for that group, but the periptychids (all *Anisonchus*) are moderate in size, much smaller than the contemporaneous *Periptychus*, and most of the carnivores are also of middle size, with *Deuterogonodon* very rare and other large carnivores absent, although they were common at this time in the Torrejon.

This rarity of large animals is still more obvious in the quarries. Phenacodonts and pantolambdids are relatively very rare, *Claenodon* is uncommon, and other large mammals do not occur. The most abundant species, *Ptilodus montanus*, ?*P. sinclairi*, *Leptacodon ladae*, *Aphronorus fraudator*, *Paromomys maturus*, *Palaechthon alticuspis*, *Metachriacus punitor*, *Didymictis microlestes*, and *Ellipsodon aquilonius*, are moderate to minute in size.

In food habits the multituberculates were rodentlike, the insectivores doubtless insectivorous (as the word is usually used, not signifying a diet composed of insects), the primates probably mainly or exclusively frugivorous, the creodonts in part omnivorous (*Claenodon*), omnivorous–carnivorous (other arctocyonids), and predaceous–carnivorous (miacids), and the condylarths probably browsing, perhaps in part frugivorous, or even partly insectivorous for the smallest forms. The known fauna apparently consumed every type of food known to have been present in the area, with the possible exception of the mollusks.

The skeletal structure is too poorly known for any of these animals to give much direct insight into their locomotion or general habitus. By analogy and comparison with allied species and genera, the multituberculates and insectivores were unguiculate and probably in good part arboreal in habitus. The primates, also, may well have been mainly or entirely arboreal. The creodonts were probably mainly terrestrial, but it is probable that some of them were at least semiarboreal. The abundant hyopsodontids were probably, judged from Torrejon and later allies, more unguiculate than ungulate in general habitus, and the smaller forms may well have been arboreal. The less common larger condylarths and the very rare pantodonts were probably terrestrial. The evidence is not at all conclusive, but it warrants the tentative conclusion that this fauna is largely arboreal, which is well in accord with the evidence that the quarries were in a swampy and heavily forested area and would go far toward explaining the unusual facies of the quarry faunas. There is, indeed, a decidedly fossorial humerus (of unknown association with teeth) in the collection; the facies association of arboreal and fossorial animals is not uncommon and is in accord with a forest environment.

The ordinal composition of the Scarritt Quarry collection differs significantly from that of the Gidley Quarry only in the almost complete absence of carnivores. Within the other orders, the family composition is as nearly similar as would be expected in deposits of similar facies but different ages except among the Primates. The abundant Gidley Quarry types, *Paromomys* and its closer allies, are not represented in the Scarritt Quarry collection, and instead of them the specialized, perhaps more strictly frugivorous, plesiadapids and carpolestids have become fairly common, although the first were uncommon and the latter very rare in the Gidley Quarry.

The Gidley Quarry is also interesting from the unusual occurrence of its fossils and the indications of the possible conditions surrounding death and burial of animals. The remains are invariably fragmentary, and with extremely rare exceptions there is no association of specimens. The bones seldom show any signs of weathering or rolling but are usually fractured, and even when they abut against wholly undisturbed matrix these fractures are clean, fresh breaks. Some further fracturing and dissociation have resulted from the compacting of the bed and development of slip planes, but for the most part these preceded fossilization. Most of the jaws have lost some teeth before burial, and many have lost all the teeth. These isolated teeth (clearly lost after death but before burial) are common in the collection. There are many bone fragments, but it is clear that the quantity of skeletal material present, even in the most fragmentary state, cannot by any means represent all the bones of the animals represented by their jaws and teeth.

The rather abundant presence of fish remains, often in articulation, and of aquatic reptile fragments and the presence of aquatic mollusks (rare in this quarry, but present), together with the nature of the sediments, suggest that the deposit was formed in sluggish water, perhaps a swampy stream course, ow-bow lake, or bayou. From the great variety of mammals present this evidently was not the site of a single or selective catastrophe, like many quarries that seem to represent quicksand or quagmire traps, but must have made a fairly complete sample of the mammals of the surrounding forest and (to a less extent) glades. Regardless of whether the mammals came here to drink, swam into the water, dropped from trees, or were occasionally washed in, it seems likely that the breaking and scattering of their bones, and perhaps commonly their deaths also, were the result of activities of the carnivorous fishes and reptiles. Such a history would probably explain the small ratio of bones to teeth (the former eaten and digested and comminuted, the latter less palatable and more resistant), the many clean breaks, lack of association, and also the common intervention of maceration, without apparent weathering (perhaps in part digestive, and otherwise subaqueous) between death and burial.

Matthew (1930) has stated that "we should expect to find in a single fossil quarry that the material of each genus represents a single ecologic niche, or, if more than one, that they are quite distinct. We should not, in other words, expect to find two or more closely related species living together at the same time, within the same area, and with the same habitat, causing their remains to be preserved together in the same quarry . . . Either there would be two or more species so widely different as to belong in obviously independent ecologic niches, or else there would be one more or less variable species." This is an application to paleontological data of the general principle summed up by Cabrera (1932) as the Law of Ecologic Incompatibility in these words: "Las formas animales afines son ecologicamente incompatibles, siendo su incompatibilidad tanto mayor, cuanto más estrecha su afinidad."

The Gidley Quarry fauna is ideally adapted to the application, on one hand, and to the exemplification and corroboration, on the other, of this law and of Matthew's remarks on the taxonomy of quarry faunas. The species present in it were certainly contemporaneous, and it is highly unlikely that any of the remains can have been brought from a point so distant as to have inhabited distinctly different areas. The general environment was probably essentially the same for all, although unquestionably it included distinct ecologic niches. It is possible that deposition extended over a period of years and that there was some seasonal or other periodic change in the species of the neighborhood, but this is purely hypothetical. It is most reasonable to conclude that all these animals did live together, at the same time, within the same area.

It is therefore to be expected that genera present in this quarry will either have only one species each or will have species not intergrading at all and reflecting structurally their pertinence to different ecologic stations in the area. With this in mind, the assumption was made in dealing with each genus that it did include (in this quarry fauna) just one species unless the contrary could be proved beyond reasonable doubt.

Matthew, in the paper cited above, and most other writers on the question of species making in paleontology have insisted on making due allowance for variation, or using for taxonomy only nonvariable characters, but they have adduced no real, objective criterion as to what "due allowance" should be, and they sometimes seem to overlook the fact that there is no such thing as a truly and completely "nonvariable" character. Not merely as mechanical, mathematical procedures but as a general system of logic and a grouping method useful both explicitly and as an implicit background for dealing with both numerical characters and attributes, the methods of statistics provide the desired means of measuring variation accurately and the necessary criterion as to whether this variation is or is not of the sort normal within a species. These tests and this logical background have been the basis for taxonomy in this study. If the specimens pertaining to one genus could not indubitably be separated into different groups, the conclusion has been that the fundamental hypothesis of one species to each genus was correct. If they necessarily had to be separated into different groups, and these groups could not be interpreted as based on nontaxonomic differences (such as age or sex), then and only then has the hypothesis been discarded.

Since this largely objective testing has intervened, it is not arguing in a circle to start the study with the assumption that Cabrera's Law applies, and then to consider the results as a test of the validity and an example of the operation of that law. (See Figs. 3–4 and 3–5.)

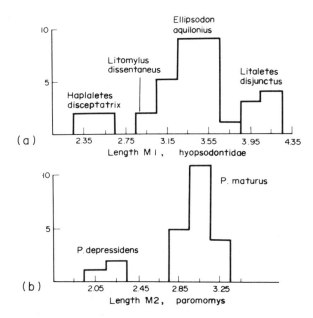

FIG. 3–4. Histograms illustrative of "Cabrera's law" as applied to the middle Paleocene mammalian fauna of the Gidley Quarry, Crazy Mountain Field, Montana. *a,* length of the first lower molar in specimens of four primitive ungulates of the order Condylarthra. In this and some other measurements the species nearly or quite intergrade. They lived in the same region at the same time, but they are morphologically quite distinct and belong to four different genera. *b,* length of the second lower molar in specimens of primitive primates. The specimens are similar and represent animals that lived in the same region at the same time, but in this dimension and other characters they do not intergrade and are classified as two species of one genus.

The actual results are as follows: Thirty genera are represented in the Gidley Quarry by one species each. Since this includes the greater part of the quarry fauna, in general it is true here that the related (congeneric) animals living at that time in this area were of the same species. The apparent exceptions belong to six genera, two of which are here given brief special consideration and another illustrated.

The multituberculates, with four species tentatively referred to *Ptilodus* and three to *Ectypodus,* are the most striking apparent exceptions. In the first place, however, the generic designations are very dubious. It is not at all certain that the species referred to *Ptilodus* or to *Ectypodus* are really congeneric in each case. There may well be one or more other genera represented here, although it does seem unlikely that each species could belong to a different genus. Aside from this possibility, the species referred to one genus are in every case sharply distinct from each other, with no intergradation shown.

Thus these species probably represent more than two genera and in any case are so sharply distinct that each must have had its own ecologic niche. They do not intergrade, but in some cases, notably *P. montanus,* there are known species, in this case *P. mediaevus,* with which they do nearly or quite intergrade but which did not live together

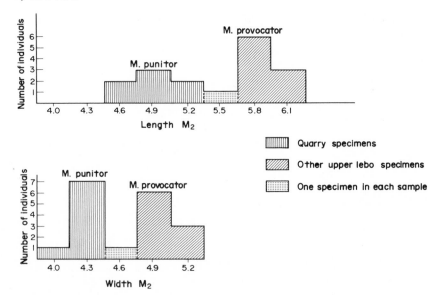

FIG. 3–5. Histograms of length and width of second lower molars of a primitive carnivore, *Metachriacus*, from the middle Paleocene of the Crazy Mountain Field, Montana, distinguished as to whether they were found in quarries or elsewhere. In these and other characters the distributions are distinctly bimodal, although they overlap to some extent. The different rock facies support the view that the two modes differ in representing animals in different environments, and they were interpreted as two species of one genus.

with them. In short, these are not exceptions to but are striking exemplifications of Cabrera's law. Fig. 3–5 illustrates another example of the same sort.

Source

(1937) The Fort Union of the Crazy Mountain Field, Montana, and its Mammalian Faunas, *U. S. Nat. Mus. Bull.* 169. (The foregoing excerpt is slightly modified from pages 59–67.)

Other References

CABRERA, A. La Incompatibilidad Ecológica: una Ley Biológica Interesante, *Anal. Soc. Sci. Argentina*, Vol. 114, pp. 243–260.

MATTHEW, W.D. (1930) Range and Limitations of Species as Seen in Fossil Mammal Faunas, *Bull. Geol. Oc. Amer.*, Vol. 41, pp. 271–274.

RAUP, D.M., and S.M. STANLEY (1971) *Principles of Paleontology*, W. H. Freeman, San Francisco. (Paleoecology, pp. 194–254, taphonomy, pp. 245–253.)

The Long View

Comment

The following text includes virtually the whole text of an ambitious, perhaps over-ambitious, contribution to a symposium on "Diversity and Stability in Ecological Systems" at Brookhaven (New York). Only some footnotes, references, and other details appropriate for a more specialized audience than is anticipated for this book have been omitted. Citations of just my main sources are still given. It may be advisable here to explain a few technical terms that are retained and not defined in the text:

Autotrophs are organisms, such as most plants, that require no organic food. Heterotrophs are those, such as all true animals, that do require such foods.

Procaryotes are very primitive organisms, like bacteria, that do not have organized nuclei and meiosis of chromosomes. Eucaryotes, like most true plants and true animals, have nuclei and meiosis.

Metazoans are the true animals, with multiple and differentiated cells, as distinct from the Protista, with single cells or undifferentiated cell clusters.

Photosynthetic organisms including some procaryotes and all the eucaryote green plants, synthesize organic compounds by means of solar energy.

Sympatric organisms live together in the same community. Allopatric ones do not.

I should add that the organic nature of the reported fossils aged about 3.2 eons has been doubted by some extreme skeptics who do not now accept as certainly organic any over about 2.2 eons.

Text

Introduction. The existence of a natural community, or ecosystem, is the result of evolution. The achievement, maintenance, and loss of stability in such a system are evolutionary processes. It is emphasized from the start that even the maintenance of stability, if or when it occurs, is a distinct and indeed an active evolutionary process, because it requires either change (that is, it is a dynamic and not a static equilibrium) or input by some positive evolutionary factors working against change. The characteristics of communities here are many and complex. Among the necessary basic data are the numbers and kinds of large and small taxa in the community. Plainly also involved, among other things, are the niche of each taxon, interactions between taxa, and their connections in the ecosystem overall, for example in food chains or webs. Changes in systematics, intricately related to changes in other ecological factors, are also multiple in nature and causation. They include, for instance, changes *in situ*, including originations and extinctions of taxa, and geographic movements, immigrations and emigrations of taxa.

Any contribution that a paleontologist can make to these matters will be primarily in the field of evolutionary systematics and should refer especially to the long view, the history of life as a whole and parts of it over geologically significant spans of times. Such enquiries will be relevant to statements like the following, by one of the most eminent neoecologists (Eugene P. Odum): "The 'strategy' of succession as a short-term process is basically the same as the 'strategy' of long-term evolutionary development of the biosphere—namely, increased control of, or homeostasis with, the physical environment in the sense of achieving maximum protection from its perturbations."

Discussion will be mainly in terms of systematics, because paleontological data, although by no means confined to this field, are best in it among those here pertinent. It will also be almost entirely in terms of the higher systematic categories, from families up to phyla. Among reasons for that, three are impelling: first, to discuss more than three billion years in less than an hour, the terms of discussion must be broad; second, stabilization and change at high levels are correspondingly highly important and are directly approached only by paleoecology; and third, paleoecology can indeed provide examples of community evolution at low systematic levels, but at present they are scattered and mostly isolated. They derive more from neoecology than they contribute to it.

Precambrian Communities. One principle of research probably comes nearer than any other to being without exceptions; no matter what your problem is, there are not enough data to solve it. That applies with bitter force to paleoecology. No complete fossil record exists for any region on earth or for any extended group of organisms. The extent of the fossil record and the continuing acceleration of our knowledge of it are quite extraordinary, but the record that can ever be obtained is woefully incomplete. Not even a single biota at any place or any time in the past can ever be completely known. Paleoecologists do find some comfort in the fact that for no community on earth today do neoecologists know the entire living biota, and yet this ignorance has not stopped them.

The paucity of fossils from the Precambrian has long been notorious. It has been emphasized by the fact that many claimed records have been so dubious as to their organic nature, their Precambrian age, or both, that they should not be taken seriously. The commonest of Precambrian fossils are laminated calcareous bodies known as stromatolites, varying from small isolated masses to whole reefs. They occur from two eons b.p. and possibly even earlier. (An eon is a billion years, American style, that is 1000 million, and b.p. is geochronological style for "before present.") It is now established that most stromatolites were formed by the trapping of sediments in algal mats. These mats were taxonomically diverse and commonly multispecific, but seem for the most part to have been formed by so-called blue-green algae (Cyanophyta).

A remarkable development of recent years has been the discovery of microscopic fossils of increasingly great Precambrian age. At present the best known, most varied, and most certainly organic of these very ancient fossils are those recorded by Tyler and Barghoorn from the Gunflint chert of Ontario, Canada. Some of these organisms can hardly be distinguished morphologically from primitive Recent algae, again Cyanophyta. Rod-shaped and coccoid bacteria, somewhat like certain recent iron bacteria but necessarily of highly uncertain physiological and ecological character, also occur. Radiometric datings have varied from 1.6 to 1.9 eons and perhaps the lately popular approximation of some two billion years is warranted.

More recently bacterium-like objects, only about one-half micron in length, moderately larger fibrillar material, alga-like spheroids, and hydrocarbons such as are now commonly of biogenic origin have been found in the Fig Tree series of South Africa, with a minimum radiometric dating of 3.1 eons. In the same general geological section but in the older Onverwacht series, with a minimal radiometric date of 3.2 eons, what the authors call "alga-like" and "lifelike" spheroids, 5 to 25 in diameter, have been found, again with possibly biogenic carbon compounds. The evidence is impressive, but one must agree that, "Although the Fig Tree and Onverwacht organic spheroids and filaments are probably of biological origin, skepticism about this sort of evidence of early Precambrian life is appropriate." Skepticism is even more in order for reports of biogenic stromatolites as old as three eons in South Africa, but this, too, is possible.

I have now mentioned the crucial boundary cases. These and others show that certainly for two eons and with high probability for three eons cellular organisms have been abundant on earth. Most or, as far as the record shows, all of those organisms were probably procaryotic like their modern analogues, or representatives, bacteria and blue-green algae. That is, they had reached cellular organization, hence were living units in the full, unequivocal sense of the word, but had not yet evolved the nuclear-chromo-somal-mitotic apparatus of heredity, growth, and control characteristic of all present true organisms except the bacteria and cyanophytes. The stromatolite-forming organisms were almost certainly photosynthetic autotrophs, and many or most of the other cyanophyte-like microorganisms probably were. Chemosynthesis probably also occurred, but on a smaller scale and of somewhat later origin. It is now believed that life originated in the form of procaryotic heterotrophs. The decomposers in the oldest known biotas were probably still predominantly primary heterotrophs.

We thus have a picture of about the simplest possible ecosystems capable of indefinite continuation, with only two trophic levels: composers (photosynthetic autotrophs) building up organic structure and materials, and decomposers (mostly or entirely simple, primary heterotrophs) breaking down and recycling. The picture is somewhat but not greatly complicated by the fact that both groups seem to have been already taxonomically somewhat diverse as early as we know or can infer them. There was already some niche specialization, perhaps more biochemical than otherwise environmental, but of a limited and comparatively simple kind. The next major evolutionary change would have been the development of protistan and algal eucaryotes. That would greatly increase the possibilities for niche specialization and phyletic adaptability, but would not in itself necessarily produce trophic relationships much more complex than those of the two-level ecosystems. Eucaryotes are definitely known from the Precambrian, and they must have originated long before, but there is little hard evidence. On hypothetical grounds, Cloud places the event toward two eons before present.

The next major event, from an ecological point of view, was the rise of metazoans or true animals. That, too, was certainly Precambrian but is very poorly recorded. The reason is, at least in part, that certainly most and possibly all Precambrian metazoans were soft-bodied and therefore preserved as fossils only under exceptional conditions. Numerous apparent trails, casts, and varied enigmatic markings in Precambrian rocks have been claimed as evidence of metazoans, but remain too dubious for definite acceptance. The oldest known unquestionable metazoans are those of the Ediacara fauna of South Australia, with similar but less rich occurrences elsewhere in Australia, in South Africa, and in England. The Precambrian age of these fossils has been denied, but this seems to be largely a formal question as to where the conventional line between Precambrian and Cambrian should be drawn. These fossils are with considerable probability somewhat but not greatly older than the long-known and classical faunas universally recognized as early Cambrian. Their age may be on the order of 0.7 eon (700 million years). Multicellular, eucaryotic algae of approximately the same age are known, but no higher plants.

The Ediacara fauna includes quite definite Coelenterata and Annelida along with other metazoans that cannot confidently be referred to any otherwise known phylum. All are soft-bodied, and it is now generally agreed that the main reason for the scarcity of Precambrian and abundance of later animal fossils is that only around the beginning of the Cambrian did metazoans commonly acquire hard parts. Many reasons for this change have been suggested, but for the present topic that is largely a side issue and cannot be pursued here. Other major groups that appear in the earliest Cambrian almost certainly were present in soft-bodied form in the late Precambrian seas: Archaeocyatha (sponge-like forms confined to the Cambrian), Porifera (true sponges), Brachiopoda, Mollusca, Arthropoda, and Echinodermata. Varied as those groups are, their earliest representatives were ecologically much less varied than they are today or, indeed, as we shall see, than they became not long, geologically speaking, after the Cambrian. They constituted essentially a single trophic group, all being basic secondary heterotrophs feeding on small to microscopic plankton (so-called nannoplankton and hekistoplankton) or scavenging likewise small organic particles. Trophic levels and many ecological roles soon to become conspicuous and most familiar to us today, for example those of herbivores and carnivores in the usual senses, had not yet evolved.

Some students believe that the evolution of communities of that late Precambrian type from the older, still simpler Precambrian kind was quite rapid in geological terms and that metazoans do not appear much earlier in the record for the simple reason that they did not exist. That is based in part on negative evidence, with rejection of some possible positive evidence, and also in part on ingenious but still quite speculative views on atmospheric evolution, which even if established would bear only indirectly on the point here at issue.

Whenever the latest Precambrian—earliest Cambrian kind of community evolved, it definitely existed by a time not later than 600 million years ago, and that is a datum point. Less definitely but with what looks like sufficient probability, much simpler communities, procaryotic, with only two trophic levels, but cellular and with modest but real niche differentiation, already existed approximately 2½ eons earlier. Communities did not in fact stabilize at either of those levels. There was a distinct increase in ecological complexity and probably also in biomass and energy flow among known Precambrian communities. The overall rate of those changes was evidently extremely slow. What has been summarized up to this point is further summed up in extremely diagrammatic form in Fig. 3–6.

Aquatic Animals. The Cambrian period is believed to have spanned approximately 100 million years. During that long time numerous major groups of aquatic animals appeared one after another in the fossil record. This phenomenon indicates not only progressive evolution but also, with high probability, the acquisition of increasingly hard, more readily fossilizable tissues by groups that had originated but not been fossilized earlier. By the end of the Cambrian all but one of the phyla that include readily fossilizable taxa are present in the fossil record. The exception, Vertebrata, appears in the early Ordovician.

The total fossil record of aquatic metazoans is summarized in Fig. 3–7 in terms of the numbers of orders known for each readily fossilizable phylum in each of the geological periods. Increase in number of higher taxa, here phyla and orders, is assumed to indicate, at least roughly, increase in the ecological complexity of particular communities and also in the diversity of existing communities. Orders as they arise usually represent distinct

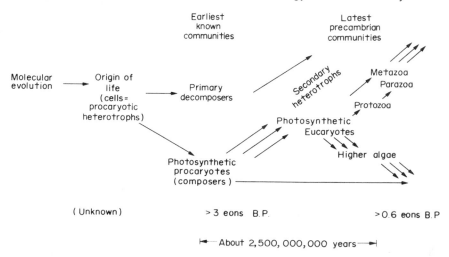

FIG. 3–6. A theory of the sequence and evolution of ecosystems in the first three or four eons (billions of years) after the origin of life. Time is scaled approximately from left to right. Arrows are simplified representations of lines of organic descent. (For fuller explanation see text.)

adaptive types, which become distributed more precisely into niches at lower taxonomic levels. The distinction of adaptive types, therefore of ecological roles, broader or narrower in approximate correlation with categorical levels, tends to become greater or more precise as the groups concerned become sympatric. Darwin was well aware of this phenomenon and discussed it clearly in *The Origin of Species*. It has more recently been rediscovered and is now sometimes called "character displacement" with examples at low taxonomic levels.

On the other hand, allopatric groups may be similar or virtually identical in ecological roles and yet taxonomically quite distinct. In such cases a large number of taxa, overall, does not correlate well with ecological complexity or the number of ecological roles. That is particularly true at low categorical levels. For example, the fact that among South America, Africa, and Australia no two have any species of native mammals in common obviously does not mean that their communities have no ecological resemblances. That sort of duplication may occur even at levels as high as orders. For example, both North and South America have had abundant, ecologically similar ungulates throughout the Cenozoic, but for most of that time they had no orders of ungulates in common. However, the great majority of aquatic, and especially marine, orders are widespread, and among these groups zoogeographic distinctions and ecological duplication are usually at familial or especially at generic, specific, or even lower categorical levels. That apparently was already true in the Paleozoic. Therefore it is reasonable to believe that in following the numbers of orders in the fossil record we are indeed following the approximate overall course of ecological complication and diversification even though in a necessarily loose way.

The ordinal totals, as seen in Fig. 3–7, rose rather steadily throughout the Cambrian and into the Ordovician. The totals are for whole periods, and both Cambrian and Ordovician were exceptionally long periods; a curve with more frequent datum points would have a lower start and a longer climb. Thereafter into the Triassic there is a

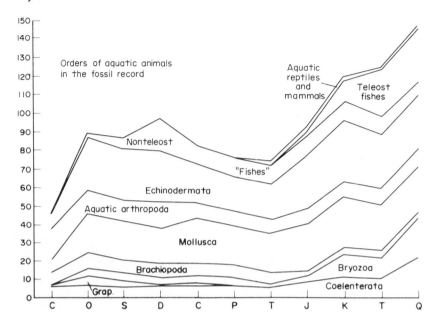

FIG. 3-7. The record of readily fossilizable groups of aquatic true animals (Metazoa) from the Cambrian to the Quaternary (present) periods. The periods, with customary abbreviated symbols, run from left (oldest) to right sequentially without being scaled to their lengths in years. For each period the number or orders within each named group is scaled on the ordinate. "Grap." is an abbreviation for graptolites, an extinct Paleozoic group of uncertain affinities.

gradual drop, largely chargeable to the echinoderms and perhaps partly an artifice of split classification for early echinoderms. It is compensated to some extent from Devonian onward by the nonteleost "fishes" (used broadly to include Agnatha, Placodermi, Chondrichthyes, and nonteleost Osteichthyes of Romer's classification). The Triassic low point is, however, a generally recognized feature of the fossil record, reflecting the major extinctions that occurred around the Permian–Triassic boundary and the faunal turnover that was used early in the history of geology to set that boundary, which is the Paleozoic–Mesozoic boundary. The extinctions and turnover may reasonably be imputed to geographic and climatic factors, progressive uplift of continents, withdrawal of epeiric seas, and deteriorating climates.

The subsequent buildup of invertebrates and nonteleost "fishes" into the Cretaceous, nearly sustained in the Tertiary, may then be a return to the ecological norm of the Paleozoic. The upturn in the Quaternary is a defect of the record, and should be ignored for present purposes. It is certain that no orders originated in the Quaternary. This upturn is wholly due to Recent orders of bryozoans and coelenterates that must have arisen before but are not known as fossils, mainly because they are relatively soft-bodied. The Cretaceous–Tertiary return to approximately the Ordovician–Devonian level is a sort of stabilization of ordinal numbers, suggesting that there is an approximate number of orders of aquatic animals that is "right" in the sense of filling the usually available ecological situations. The Permian–Triassic episode indicates a failure to achieve maximum protection from the perturbations of the physical environment, in Odum's terms but contrary to his expectations earlier quoted. Still the subsequent recovery might be considered a very large, very long-range instance of homeostatis.

As always, there are complications, and the total picture is not so simple. Two further points are particularly important. One, striking in Fig. 3–7, is that with the teleost fishes something new was added, carrying ordinal diversity far beyond any previous level. This was not a replacement of the nonteleosts, which were as varied in the Tertiary, when there were multitudes of teleosts, as in the Carboniferous, when there were none. It seems that the teleost expansion must represent both the exploitation of new ecological situations and the greater niche-subdivision of old. The second important complication is that in most of the phyla there was major replacement of taxa even without notable changes in their numbers. The mid-Paleozoic and mid-Mesozoic orders are largely different. Examples will be given later.

Land and Air. Except for the teleost phenomenon, the principal acquatic environments seem to have been ecologically saturated by Devonian times. The next great episode in the history of ecosystems was the origin of totally new kinds of communities, those of the land, now most familiar to most of us. The first clear evidence of these occurs in the Silurian, they were more definitely developed in the Devonian, and they reached a sort of plateau of ordinal numbers in the Permian. These communities were, and are, characterized by highly varied vascular plants, insects, numerous other nonaquatic arthropods, and amphibious to fully terrestrial vertebrates. They also include some mollusks, although no order is completely terrestrial, and innumerable organisms not fossilizable or so little so as not to provide worthwhile data for the present enquiry. The data considered worth while are summarized in Fig. 3–8.

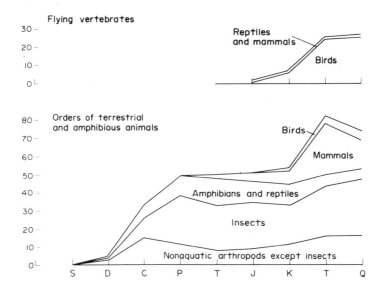

FIG. 3–8. The record of readily fossilizable groups of amphibious and terrestrial true animals (metazoans). Construction and conventions are as in Fig. 3–7.

Here another complication merits attention. The total variety of amphibians and reptiles remains essentially level from Permian to Cretaceous, but again there are radical changes in the actual taxa present and there are also changes in ecological facies. In the Paleozoic many, probably most, of the amphibians were essentially aquatic in fresh water and perhaps should be in Fig. 3-7 rather than here, although few of the orders in Fig. 3-7 are confined to fresh water. Some of the Paleozoic reptiles of what do seem properly to be considered terrestrial orders were probably virtually amphibious. (Also the later turtles and crocodiles are here counted as terrestrial to amphibious, only a few of them being fully aquatic.) In the Mesozoic the amphibians developed more strictly amphibious and even quite strictly terrestrial facies, and the reptiles evolved taxa in completely terrestrial, even highly arid communities. The insects also changed greatly in overall facies with the spread of flowering plants, but without becoming notably more diverse at the ordinal level.

In this picture the mammals are somewhat analogous to the teleosts in the history of aquatic communities. It is here clearer that they are in part replacing or relaying earlier groups, here among the amphibians and reptiles. The multiplicity of mammalian orders is also affected by the fact, previously mentioned in passing, that there have been a number of allopatric orders, mostly among the ungulates, closely similar in ecology. Nevertheless, the Tertiary mammals did spread into new ecological situations and did subdivide niches to a greater degree than other terrestrial vertebrates.

The flying vertebrates, also shown in Figure 3-7, add another major adaptive type, intricately subdivisible, to the terrestrial faunas. Although the bats are morphologically so similar that they are usually classified as one order (occasionally two), they are world wide, abundant in most terrestrial communities, and almost incredibly varied in food habits and some other adaptations. In spite of usual classification in many orders (25 recognized here), the flying birds are almost as stereotyped morphologically as the bats and not much more varied adaptively. They happen to be more conspicuous, and it is true that within any one community they are likely to have varied roles, to belong, for example, to more different food chains.

The moderate increase in numbers of orders of birds and insects from Tertiary to Quaternary is again a defect of the record, here not very marked. The supposed poverty of those groups in fossils has become notorious, and it is rather surprising to find that they are in fact well represented at the ordinal level, at least. The sharp reduction in numbers of orders of terrestrial mammals from Tertiary to Quaternary is a real phenomenon, reflecting extinction of Tertiary orders. In view of arguments on causes of extinction and discussions of supposed Pleistocene and Recent "overkill," it is noteworthy that the strong decline in major groups of mammals shown in this figure was entirely pre-Pleistocene.

It is true in a way that we are now living among natural communities (what is left of them) of a kind that appeared in the Devonian and was well developed by the Permian, more or less 250 million years ago. Yet that requires a definition of "kind" that is very broad indeed. Even in the very simplest terms of the numbers of orders involved, it is hard to see any real tendency toward stabilization in Fig. 3-8.

Relay and Replacement. Groups like the teleost fishes in aquatic environments and like the mammals in terrestrial environments seem to be capable of more distinct and varied adaptations, hence more multiple division of adaptive zones and ecological roles than their predecessors. At low levels this is called stenoky, and the concept is also applicable at higher levels. In spite of this phenomenon, which may increase the number of taxa, high or low, that can exist within the same environmental conditions, there is support for the idea that under given circumstances there is an approximately fixed number of taxa that fill an environment. Once filling has occurred, which may take a long time (e.g. Cambrian–Ordovician for marine environments), the number tends to stabilize until there is some marked environmental change (e.g. Permian–Triassic) or until a new group capable of comparatively greater stenoky at various levels arises (e.g. teleosts).

Even though there is thus a measure of stabilization in numbers of taxa in particular environments, broadly defined, closer examination shows that essentially the same ecological roles may be, in fact usually are, played by quite different taxa in the course of geological time. Examples abound in all the phyla of Figs. 3–7 and 3–8. Only a few, different in kind, are mentioned here in substantiation. Fig. 3–9 summarizes an aspect of the history of the ammonites, an extinct group of mollusks occupying one broad adaptive zone, divided into a great number of narrow generic zones and specific niches. Three times this group (order) was greatly reduced in variety, here represented by superfamilies in the very split classification favored by most current specialists: between Devonian and Carboniferous, Permian and Triassic, and Triassic and Jurassic. Each time a small number of taxa, only one or two as indicated in this superfamily arrangement, did survive, but in each case the survivors subsequently diversified until approximately the same level of taxonomic, hence probably also of ecological diversification was regained. Thus there is a sequence of relays within the group itself, with homeostatic return to relative stability (in terms of ecological diversity) after each major disturbance. Finally, in the late Cretaceous, reduction otherwise similar to the three earlier instances did not leave a surviving stock, and any ecological relay that occurred necessarily came from other groups. (Fig. 3–9 does not have a measured time scale, and for purposes of diagrammatic simplicity superfamilies in any one period are all shown alike. Originations did not occur just at the beginnings of periods nor extinctions just at their ends.)

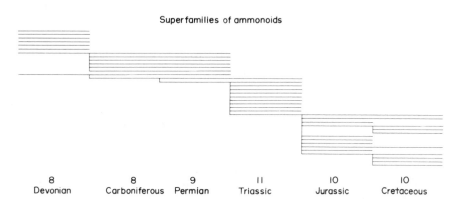

Superfamilies of ammonoids

| 8 | 8 | 9 | 11 | 10 | 10 |
| Devonian | Carboniferous | Permian | Triassic | Jurassic | Cretaceous |

Fig. 3–9. A simplified and conventionalized phylogeny of the superfamilies of ammonoids, extinct relatives of the living chambered nautilus. Each superfamily is represented by a horizontal line in the period when it occurred. Time of appearance and disappearance within periods are not indicated. Vertical lines tie descendant superfamilies, below, to their ancestry, above. Numbers above the names of periods are the numbers of superfamilies in that period. They remain essentially constant by periods despite the three evident mass extinctions, up to the final total extinction. The abscissal scale is sequential, not metric, and there is no ordinate scale.

An even more striking example of relay still within a single group, to be sure, but at higher taxonomic levels and with a different biological basis is given in Fig. 3–10. Data are for families within five orders of coelenterates, roughly speaking corals, most of which are reef-building (hermatypic), although some non-reef-building (ahermatypic) genera and species are included. Reef communities, which of course always comprise many organisms in addition to coelenterates, have such well-defined and generally similar characteristics that stabilization at some level of diversity and some kind of taxonomic composition might well be expected, but clearly it has never occurred. The number of families of hermatypic coelenterates has fluctuated continuously and greatly, with a Triassic low coinciding with that previously noted for aquatic animals in general. Only one of the three hermatypic Paleozoic orders survived into the Mesozoic, and it did not reach the Cenozoic. Moreover, it now seems quite improbable that either of the two Cenozoic orders was derived from any of the three Paleozoic orders. It is probable that both the Octocorallia and Scleractinia evolved from soft-bodied Paleozoic ancestors that began to secrete heavy skeletons only in the Mesozoic. It is a reasonable speculation that in so evoling they were moving into an adaptive zone made empty by the extinction of the Rugosa and Tabulata. That sort of opportunistic reoccupation of ecological roles might be considered homeostatic in some sense of the word, but it can hardly be called stabilization, especially as the diversification and taxonomy of hermatypic coelenterates continued to change markedly.

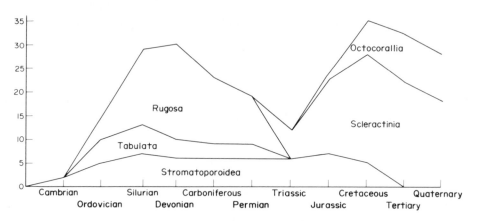

FIG. 3–10. Numbers of known predominantly reef-building (hermatypic) families of coelenterates ("corals") within each of the five named orders in each period named. Construction of the graph is as in Figs. 3–7 and 3–8.

In land communities the evolution of ecosystems, including the universal phenomenon of relay and replacement, is reflected in the changing systematics and ecological makeup of Cenozoic mammalian faunas, for which there are particularly good data. I examined changing faunal compositions in North and South America long since, and although the statistical data and distributions need updating, the general patterns are still valid. Cooke has recently analyzed the African faunas in the same way. Oddly enough, the European and Asiatic faunas have not been studied in this way, although the essential data are ready to hand. Data for Australia, although now accruing rapidly, are still insufficient.

The patterns for Africa, North America, and South America are not without resemblances but also have great differences. The African pattern is the simplest, perhaps in part because African mammals older than the late Eocene are almost unknown. With a single exception (Embrithopoda, with only one known genus) all the orders known from the Tertiary, late Eocene onward, survive today in Africa. Their importance has varied greatly, for example the Hyracoidea were abundant and adaptively varied in the Oligocene but have steadily waned to their present three genera of essentially two, not extremely different adaptive types. On the other hand, the artiodactyls became steadily more diverse into the Pleistocene and are still among the most striking African mammals, exceeded in diversity only by the rodents. Speaking very broadly, a fauna more or less of Tertiary type does survive (perhaps one should say "barely") in Africa, but at lower taxonomic levels the changes, both taxonomic and ecological, are enormous. Of 31 families definitely recognized in the Miocene of Africa, only 20 survive there, and the Pleistocene fauna was so dominated by strange, often monstrous, extinct genera and species that it would look almost totally exotic to an African of today.

In North America there has been a constant, highly complex interplay between the evolution of native stocks *in situ*, with the extinction of many of them, and the appearance of immigrants from Eurasia periodically throughout the Cenozoic and from South America in the late Cenozoic. The magnitude of the changes can be judged by the fact that of the 18 orders of mammals now definitely identified from the Eocene of temperate North America, only 8 were still here to greet man when he arrived in the late Pleistocene, and of those two (Marsupialia and Edentata) had in fact become extinct here and reappeared only in the Pleistocene as migrants from South America.

From one point of view it is not surprising that the makeup of North American communities including mammals was in a state of constant flux, given the frequent irruptions of migrants into them. However, such irruptions are normal environmental changes throughout the history of life, and the communities obviously were not stabilized in the sense of being buffered against them. That is particularly striking when one realizes that many of the ecologically most important mammals in our present natural communities are newcomers and have greatly changed the nature of those communities. Among them are bison, deer, wolves, brown and grizzly bears, numerous rodents, and many more.

It would be particularly interesting to know what would happen from internal causes in an ecosystem that remained closed, that is, that underwent no environmental changes from the movement of other taxa into it from outside. It is unlikely that any ecosystem ever did remain wholly closed in that sense for significantly long periods of time. However, terrestrial communities in southern South America probably remained as nearly so as any in late Paleocene through Eocene and again through Miocene and early Pliocene, and for those times we have a reasonably good record of the mammals in those communities. As would be expected, the general characteristics of known faunas did not change much within either one of those spans, but there were many changes more or less of detail. No genus is common to the probable late Paleocene (Riochican) and probable late Eocene Mustersan). Of 14 known late Paleocene families, 11 were still present in the late Eocene, and at the latter time there were 7 known families apparently not yet present in the Paleocene. That may in part be due to deficient knowledge of the Paleocene, but several of the later families do seem to be of truly later origin, and in almost all families common to the two there are definite advances or specializations.

Between the Eocene and the Miocene two important orders new to South America entered the continent, rodents and primates, and they gave the later faunas a very different aspect. Within the Miocene and early Pliocene changes again are less profound, but again are quite definite, with no genus spanning that whole time and with all families showing progressive changes. Again the families on the whole remained the same, but a few old families became extinct and a few new ones, among the lately immigrant Rodentia, arose.

The greatest change in South American faunas occurred after land connection with North America was formed or resumed, mainly in the Pleistocene. There was then great faunal interchange between the two continents, one interesting aspect of which is diagrammatically summarized in Fig. 3–11. Before the interchange began there were no families of terrestrial mammals in common between the two continents. During the interchange each continent received a number of families from the other, and despite some extinctions the number of families rose markedly on each continent. Subsequently extinctions, which affected families of southern origin much more than those of northern origin (another point not pursued here), the number dropped on both continents approximately to the numerical level that obtained before the faunal mixture. This strongly suggests, or at the very least is consistent with, the idea that each continent was ecologically full of land mammals before interchange and that the number of about 25 families in North, about 30 in South America represents ecological saturation. The top condition in the Pleistocene would then represent supersaturation and the later drop, a return to normal. An interestingly similar point as to whether Africa now has as many birds as "belong" there for ecological saturation has been raised lately in a book by Moreau and a review of it by Martin.

Among many further points tempting discussion, just one more cannot be resisted. It could be argued that South America has an ecological deficiency because it lost many more ungulates by extinction than it gained by immigration and subsequent survival. It does now have an apparent paucity of grazing and to less extent of browsing ungulates.

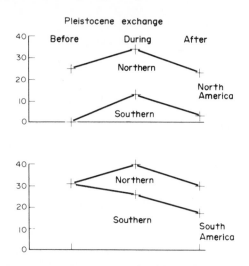

Fig. 3–11. Numbers of families of land mammals involved in the Great Intercontinental Interchange between North and South America in the late Cenozoic. The ordinate scale indicates numbers of families in nontropical North America, above, and in South America before the interchange, at the height of interchange, and at present after the interchange.

The same could be said of temperate North America, which now has a much smaller variety of ungulates, only one order and only ten species, than at any other time in the Cenozoic. Far the greatest number of extinctions bringing about this dearth on each of the continents occurred long before the advent of man on either. On both continents the waning of the ungulates has occurred pari passu with the waxing of herbivorous rodents. It is at least a reasonable hypothesis that these were also ecological replacements, which do not have to involve superficially similar organisms.

Conclusion. Brevity here suggests a single conclusion: if indeed the earth's ecosystems are tending toward long-range stabilization or static equilibrium, three billion years has been too short a time to reach that condition.

Source

(1969) The First Three Billion Years of Community Evolution, *Brookhaven Symposia in Biology*, No. 22, pp. 162–177.

Other References

MOORE, R.C., editor. (1957 and continuing) *Treatise on Invertebrate Paleontology*, Geological Society of America and University of Kansas Press. (Source of most of the data on fossil invertebrates in this paper.)

ROMER, A.S. (1966) *Vertebrate Paleontology*, University of Chicago Press. (Source of some data on vertebrates other than mammals; other data on vertebrates were my own.)

SCHOPF, J.W. (1975) Precambrian Paleobiology: Problems and Perspectives, *Ann. Rev. Earth and Planetary Sci.*, Vol. 3, pp. 213–249. (An updating; it does not seem to me to require much change in this part of my 1969 paper.)

4

Systematics and Taxonomy

PHYLOGENY and classification are involved to some extent in all considerations of evolutionary and organismal biology. To the many biologists engaged in the history of life and in the classification of organisms they may be the principal or even the only topics. Among such biologists the paleontologists are particularly concerned with these subjects, not only in their application to a given fauna, flora, or taxon (classificatory group) but also in attention to the principles, in one sense of the word the philosophy, involved. It is the latter aspects that are especially considered in this present collection of previous essays or excerpts from them.

Systematics is the study of the diversity of life, past and present, and of relationships of any and all sorts among organisms. Taxonomy, although often equated with classification, is better defined as a sort of metaclassification in a distinction analogous to that between metaphysics and the natural sciences (physics in a classical sense): it is the study of the principles and philosophy of classification. Phylogeny involves one sort of relationship among organisms, that of evolutionary descent. Thus I see phylogeny and classification as two special subjects within the broader scope of systematics. What congruence or relevance is considered as existing between phylogeny and classification depends on one's views about taxonomy. Here there is a special point that should be briefly discussed and as far as possible cleared up before proceeding with the separate topics of the rest of this section of the book.

The important point here is that phylogeny and classification are two things, not only separate but also different in kind. Most taxonomists will concede that as obvious, but in their work there is often a confusion of the two or, at best, a lack of logical clarity in their treatment of the relationship between them. The fundamental difference is that phylogeny is something that happened and classification is an arrangement of its results. Although phylogeny cannot be observed as such over periods long enough to be really significant, it existed as a sequence of factual events among real things (organisms) and in a philosophical or logical sense it is objective or realistic in nature. Classification is not. It is an artifice with no objective reality. It arises and exists only in the minds of its devisers, learners, and users. There is an aspect of classification as an art and of the taxonomist as an artisan, which does not mean that classification is not, or at least that it cannot be, also a science and the taxonomist a scientist.

I am here speaking of classification in the way the word is customarily used by organismal biologists, as a formal hierarchic grouping of organisms according to relationships among them and a naming of the groups so formed. It is true that at least a partial arrangement meeting that definition might be objective, for instance one grouping together the animals observed to be aquatic, and something like this is done by ecologists,

110

but that is trivial for the present topic. The ecologists themselves do not call such groupings zoological or botanical classifications.

It is possible to make a zoological or botanical classification with no reference to phylogeny. In fact that was done for generations and in quite workable ways by pre-evolutionary naturalists. There is even a modern school of taxonomists, those who first called themselves "numerical taxonomists" but are more accurately designated as pheneticists, who as a point of principle intended to exclude phylogeny from consideration in classification. Some of them even called this "objective," but the only truly objective element back of the relationships they used in classification was in fact phylogenetic. Some of them have come around to this point of view and draw phylogenetic conclusions by means of their sometimes quite elegant methods.

At another pole is a school that purports to base classification exclusively on what some of its exponents believe to be objectively recreated phylogeny. This school describes itself as phylogenetic but it is not the only or the majority group of taxonomists who relate classification to phylogeny, and it is more clearly identified as Hennigian, after the late Willi Hennig who was the leading expounder of those views. This is not the place to summarize the Hennigian procedures beyond pointing out that they produce visual treelike diagrams, dendrograms, on which their classifications are based. These, like all classifications, are based on observed characteristics of organisms and they are useful keys to distinctive characteristics. As some of their formulators have come to recognize, these dendrograms are not unique, that is, different students using the same facts and following Hennig's procedures may come up with different dendrograms. It is also becoming evident that while these may approximate phylogenies, they do not necessarily—or perhaps usually—do so, and some of their applications to classification are at best cumbersome, at worst illogical, contradictory, and impractical.

My own, more middle-of-the-road views agree in general principles with those that have always been and I believe still are held by most working evolutionary taxonomists, of course with minor differences of opinion and often practiced without much attention to abstraction of principles. This school—to call it that—used to be called "evolutionary" or "phylogenetic," but all competent biologists are new evolutionists and the term "phylogenetic" has become confused with "Hennigian." I now prefer to call this approach "eclectic," for indeed in the field of taxonomy it does seek to try all things and hold fast to that which is good. Some of the aspects of phylogenetic principles and of eclectic taxonomic theory are brought out in the other parts of this section of the book and some of them more fully in the following previous books:

(1944) *Tempo and Mode in Evolution*, Columbia University Press, New York.
(1945) The Principles of Classification and a Classification of Mammals, *Bulletin of the American Museum of Natural History*, Volume 85.
(1949) *The Meaning of Evolution*, Yale University Press, New Haven and London.
(1953) *The Major Features of Evolution*, Columbia University Press, New York.
(1961) *Principles of Animal Taxonomy*, Columbia University Press, New York
(1967) *The Meaning of Evolution*, Revised Edition, Yale University Press, New Haven and London.

Patterns of Cladistic Evolution

Comments

The essay here reprinted no longer seems fresh because most of its concepts, unorthodox when it was written, have now become usual even to the point of being taken for granted. Nevertheless I find inclusion of this text here warranted because its ideas still seem sound, with some slight re-expression, and an early statement of them has some special interest. There is also a personal reason that may be justifiable. When I wrote this I had already been working at phylogenetic and taxonomic problems for more than a decade but only at this time did I feel confident and experienced enough to attempt more general expositions of some principles involved. It marked my abandonment of the typological thinking of my college teachers and started aiming me toward statistical biometry and the deeper investigation of evolutionary theory and taxonomic stance.

A few words here and there have been changed to make this essay more readily comprehensible to present students. In this text the term "phylum" is usually applied to a specific line of descent in a sequence of populations, which is now usually and here occasionally called a "clade". However the term "phylum" has been left, with the meaning of "clade," in the figures.

It should be noted that although the splitting of clades is represented by known fossils in some cases, these examples are not as frequent as I expected them to be when I wrote this. The most likely reason for this is not that such events were not the usual course of speciation in the sense of proliferation of clades but that the origin of new specific branches from continuing parental species has usually occurred by the isolation of small marginal populations. The odds are strongly against the preservation and discovery of a clear and complete fossil record of the whole event.

Text

In the fine optimistic years of early evolutionary paleontology, it became fashionable to draw up genealogical trees for various groups of animals, and especially of mammals, in which the supposed interrelationships of all these creatures were shown by representing the living animals as the ends of twigs and their ancestors through preceding ages as lower and connected parts of the stems, branches, and trunk of the tree of life. It was generally recognized that certain details remained to be filled in, but on the whole the concept was simple and satisfying, and much of the work of that period gives the impression that the missing details were not considered important and that their discovery was considered imminent.

The use of such graphic means of indicating descent is still common, and there are few paleontologists who have not constructed some sort of phyletic tree, either in words or in an actual diagram. However, a cynical note has crept in, and phylogeny has been subjected to severe criticism, not only by our colleagues, the biologists, but also among paleontologists themselves. As almost every family tree proposed by our predecessors proved, beyond much doubt, to be erroneous in essential features, it is not surprising that much of the old whole-hearted faith in the imminent solution of all such problems has now been lost.

It is particularly striking that the old picture of phylogeny as a branching tree has almost entirely disappeared from detailed monographs of special groups and that in its stead appear diagrams suggesting many different lines that are parallel or that diverge but are not connected. A single one of the innumerable instances in recent literature is the phylogeny of the mastodonts as given by Osborn. It has fifteen subfamilies and suggests at least thirty distinct major lines of descent, or independent phyla. These phyla are essentially disconnected and, as far as that eminent authority could determine, none was clearly derived from any other. This example is not the most extreme, nor do I mean to suggest that it is necessarily erroneous, but only that it typifies the tendency toward multilinear interpretations. If many of the recent odontological and taxonomic monographs of special groups of mammals are studied, the impression is given that no two mammals, differing in any obvious or significant respect, ever actually had a common ancestor, but that each restricted little phylum has been distinct since the beginning of time. One is almost forced to think that when primordial cells first arose on the planet there were already potentially man-cells, oyster-cells, codfish-cells, ape-cells, and every other distinct sort to correspond to every minor phylum of living things that ever was to arise, and that these various blobs of protoplasm were not on speaking terms. Of course, no one really believes this, and in theory all subscribe to a more rational evolutionary picture, but in practice the common ancestry of two or more phyla is, more often than not, simply listed as hypothetical or undiscovered, and indeed seems to be used more as lip-service to the evolutionary principle than from any deep conviction as to its reality.

These new pictures with their multitudes of independent lines are unquestionably nearer the truth than were the naïve trees formerly in vogue. If the tree symbol has any validity at all, there were incomparably more branches than was supposed even a few years ago, and the tree was one of most peculiar habit, most of the branches springing from near the ground and then, independent of each other, rising to great heights. On the other hand, it seems that this new conception is now so well established that it may become a habit of mind or a deep prejudice and that the facts will be fitted to it instead of its being adapted to the facts as they are discovered. From the nature of the case, it is well recognized that the chance of our finding the direct, or even the close structural, ancestor of any given species, genus, or other group is small, but it is inconceivable in any present theory as to mammalian history that this chance should not be large enough to be a real factor in our studies. As hundreds and thousands of new fossil mammals are discovered and such a high proportion of them are announced as representing new phyla without known ancestors or descendants, there comes the uneasy feeling that perhaps the pendulum of scientific fashion has swung too far. Certainly we still believe in descent with modification; yet with two really different animals, one older and one younger, before us, it has become difficult to convince ourselves or our colleagues that in this concrete instance one is actually derived from the other.

This uneasy feeling about the mass psychology of the followers of our science is accentuated when the methods by which these conclusions are reached and the arguments by which they are sustained are scrutinized. Here I may appear in the unfavorable light of a Daniel come to judgment without the precaution of bringing any credentials; I can only say that my criticisms are directed as much against my own work as against that of anyone else, and that I am not satisfied with my progress even in this rather personal field, although I can hope that I have achieved some improvement. In considering the methods used by any student of phylogeny, the natural questions are whether he has really considered all the possible interpretations (consistent with the generally conceded premises of evolutionary paleontology) and whether he has any impelling, impersonal, or objective reason for supporting some particular one of them. As regards an appreciable amount of recent phylogenetic research, the only honest answer to both these questions is "no."

In order to examine more definitely the bearing of these general considerations on problems of phylogenetic research, several hypothetical and generalized but typical examples of such research may be considered. In the first place, take the familiar, routine task of classifying a number of similar specimens from a limited horizon and locality (Fig. 4–1). Although postulated as similar, it is certain that their characters cannot be identical, for, as far as has ever been discovered or can be conceived, no two animals identical in character ever existed. There is, then, a range of difference in their morphologic characters, which may be symbolically represented by making a cross for each specimen and placing the crosses at different points on a single horizontal line, although, of course, the differences will really be more complex than such a diagram suggests. It is axiomatic on the basis of current evolutionary philosophy that these individuals are samples from, and that the line on which they are arranged in a cross section of, one or more phyla or lines of descent. The questions to be answered are: How many phyla are represented and what are their limits? Current methods, which tend to a very polyphyletic view of mammalian descent, might well lead to the recognition of three distinct groups among the specimens at hand; hence, to the conclusions that they represent three distinct phyla (Fig. 4–1A). The specimens themselves would probably be placed in three species. In the diagram there is no obvious three-fold division, and the lines between the species are placed more or less arbitrarily so as to give each supposed species about the same scope. This is not an unfair representation, for every taxonomist knows that exactly such a division has been made time after time and still is being made daily—for instance, between small-, medium-, and large-sized individuals, all of which intergrade to some extent. On the other hand, the whole group of specimens may belong

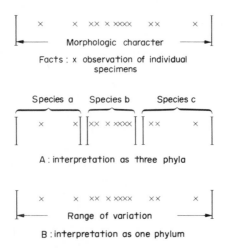

FIG. 4–1. Several similar but varying specimens from one horizon and locality interpreted as parts of three lineages or clades (A) and as parts of a single lineage or clade (B).

to one race or species, and the whole difference between them may be the range of variation within an essentially unified phylum (Fig 4–1B).

In one form, this problem boils down to the old question as to what is a species (or a subspecies, or a genus). It has been endlessly discussed since Linnaean times and probably will be disputed as long as there are zoologists. I do not propose to linger long on that aspect of the matter, although it does suggest one example of absurdity that passes almost unchallenged in paleontological research. I was once taught, and I have often heard from other paleontologists or observed as tacitly accepted in their work, that if one specimen is as much as 15 per cent larger than another in any linear dimension, it is safe enough to assume that they belong to different species. The fact is, as everyone who has studied variation should know, that there is hardly any geographic race, let alone any species, in which some variants are not more than 15 per cent larger in some dimensions than are others. On the other hand, a difference in the average dimensions of two groups of individuals may be a perfectly valid and highly significant specific character, even though the difference is considerably less than 15 per cent.

The only other point to be made here with regard to this example is that, if the specimens really represent three distinct phyla, there must be some objective difference between them, not as individuals but as groups. Differences between the individual specimens do not in themselves necessarily have any phyletic significance, for two individuals always differ from each other, but, if such varying differences can be shown to characterize and to differentiate groups of individuals, they take on wide meaning. One of the most conclusive ways in which such group characters can be differentiated is by means of frequency distributions (Fig. 4–2). If the specimens really represent three different groups (here postulated as being clades), then in some characters they will almost inevitably show separation in a frequency curve with three different modes, or apices. If, on the other hand, only one group is really present, then the frequency

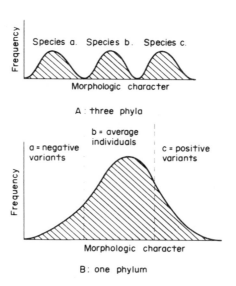

Fig. 4–2. Generalized and idealized sorts of frequency distributions of distinctive characters corresponding with interpretations (A) and (B) of the specimens diagrammatically shown in Fig. 4–1.

distributions will usually show that the supposed three groups include merely the negative, average, and positive variants within a single natural unit.

It should be emphasized that curves derived from actual data are never so smooth as those shown in the diagram, that double or multiple frequency curves do not always demonstrate phyletic separation, and that there are other difficulties and disabilities in the way of this method. Paleontological research will always require a certain amount of personal judgment; this method is not a universal solution of phyletic problems, but it is a remarkably useful tool, and its almost total neglect in this field is unfortunate.

One rather obvious difficulty is that much research on phylogeny, insofar as it uses the direct data of paleontology, must be based on few, or even on single, specimens. Some of the precise methods, many of the results, and most of the basic logic of the interpretation of frequency distributions are, nevertheless, applicable to small samples and even, strange as it may seem, to unique specimens. This is too complex a subject to discuss further here. Moreover, there are available entirely adequate and even large samples of many fossil mammals. In most institutions, only the one or two best specimens, and those that supplement them, are considered to have any significance and only they are studied with any care. The others are "duplicates," to be discarded, sold, exchanged, or forgotten. Nevertheless, it is only from these duplicates that many basic problems of phylogeny can ever be satisfactorily solved. Paleontological psychology has been to direct attention, to a disabling extent, to the individual specimen, and the number of publications on fossil mammals in which a suite of specimens has been adequately and correctly studied as representative of a group can almost be counted on the fingers of one hand.

A second example more nearly approximates the usual conditions of phylogenetic research on paleontological materials (Fig. 4–3). As before, morphologic differences are symbolized by the horizontal distribution of crosses, representing individual specimens, whereas a second, vertical dimension is introduced to represent the distribution of these specimens in time—that is, their geologic ages.

In this instance, three principal types of phyletic interpretation are possible, and all three should be considered in each case. The first general type of interpretation is to

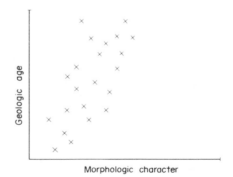

Facts : x observation of individual specimens

FIG. 4–3. Diagram of samples of similar but varying specimens extending over an appreciable span of geological time.

consider them as two or more clades, perhaps related but essentially independent—that is, not ancestral or descendant. An interpretation of this type is that exemplified as two clades, partly overlapping in time, evolving in parallel directions, as suggested by the arrows in the diagram (Fig. 4–4), but separated by certain relatively constant morphological distinctions, symbolized by their relative positions in the diagram. This is now the fashionable interpretation, and it is almost invariably applied by someone, sooner or later, to data that fall more or less into this pattern. Hence, we have the current picture of parallelism as an all but universal phenomenon in mammalian evolution.

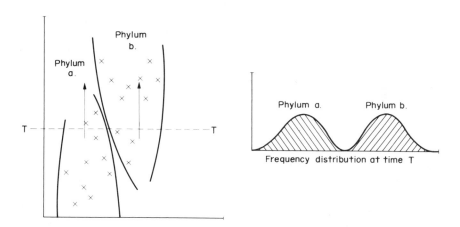

Fig. 4–4. The data of Fig. 4–3 interpreted as representing two lineages evolving more or less in parallel. The diagram to the right indicates the sort of theoretical frequency distribution expected at time *T* if this interpretation is correct.

If the criterion of frequency distribution can be applied to such cases, a sample of unified age, including parts of both phyla, should, as regards some of its characters, at least, give a curve that is significantly bimodal, or two distinct curves. For the sake of simplicity in this and other instances, these curves have been represented as approximately normal. *A priori,* several other types of curves might appear, and the problem might then be complicated even beyond what has already been suggested. In practice, however, the application of these methods in many different cases to such variates as are commonly used in establishing supposed phyletic distinctions has shown that their frequency distributions almost always approximate the normal curve closely enough to validate the use of constants and methods of analysis based on that curve.

A second type of interpretation of data of this general pattern is as a single phylum of distinctly changing character (Fig. 4–5). In this interpretation there is not supposed to be a phylum *a*, the members of which have, for example, smaller horns, and a parallel phylum *b*, with, for the sake of illustration, larger horns, but, instead, a single phylum in which an older species *a* is transformed in time into a younger species *b* with larger horns. This type of interpretation is so simple and must, in the very nature of things, have been so common in reality that it is surprising to note how seldom it is now invoked. In circumstances as simple as those of larger- and smaller-horned individuals, here employed to facilitate simplicity of verbal expression and diagrammatic exempli-

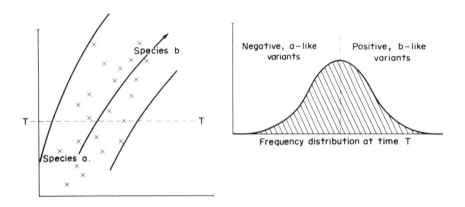

Fɪɢ. 4–5. The date of Fig. 4–3 interpreted as a single changing lineage and the sort of theoretical frequency distribution expected at time T if this interpretation is correct.

fication, this more obvious interpretation is, indeed, frequently accepted, but if the conditions are more complex, as they almost always are in dealing with concrete observations, then the polyphyletic interpretation is at least as likely to be advanced as that of simple monophyly.

If a single phylum be represented, an adequate sample unified as to time will tend to give a single and unimodal frequency curve. Here again, sight is not lost of the fact that a separation may be observed that is not phyletic but is sexual or is caused by different age groups or some other factor, but these biological considerations must in any event be kept clearly in mind and allowance for them may be taken for granted in considering the broader principles of phyletic research.

Actual examples in which a single changing phylum has been interpreted as two parallel phyla are probably common in the literature. I have encountered several cases in which the frequency distributions and other relatively objective criteria show beyond reasonable doubt that this has been done, although these criteria have, as yet, been applied in only a few of the many cases in which I hope they will eventually be used. Particularly prevalent in our literature are interpretations in which two related species (or other groups), *a* and *b*, are said to overlap in time in such a way that first *a* is abundant and *b* rare, then the two are about equally common, then *a* is rare and *b* is abundant. This, now a common type of interpretation, involving a pattern of two parallel phyla that replace each other in time by the gradual dwindling of one and increase of the other, should be viewed with profound suspicion unless it is supported by something more substantial than the subjective methods so widespread in such research. In several actual cases carefully investigated, it has been found that the facts are, almost certainly, that a single phylum is present. The average individuals of an earlier time are considered as of one phylum, *a*, and the average individuals of a later time as of another phylum, *b*. In a sample of intermediate age, variants more resembling the earlier forms are considered as of phylum *a*, and those nearer the later forms are considered as of a distinct phylum *b*, whereas, in fact, they are simply the negative and positive variants in a unified distribution, and the sample is of an intermediate stage linking *a* and *b* into a single phylum (Fig. 4–6).

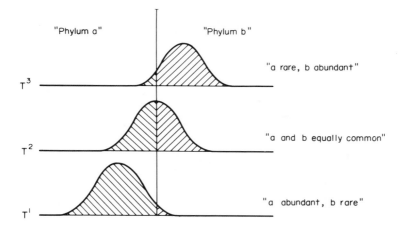

FIG. 4–6. Artificial dissection of successive theoretical frequency distributions of a single lineage erroneously attributed to two different lineages or phyla.

It is probable that in some cases the reverse of this is true and that what are really two separate and more or less parallel phyla have mistakenly been interpreted as a single phylum. Certainly, this was often true in earlier work on phylogeny, but now most of these artificially simplified phyla have been dissected; there is considerable evidence that the dissection is tending to go too far and to create many separations that did not exist in nature.

The third principal possibility in interpreting data of this general pattern is that the phyletic history is one of branching (Fig. 4–7). In such a case, an ancestral species, *a*, gives rise both to a progressive species, *b*, and to a divergent species, *c*, or else the parent stock also continues in more or less unaltered form so that the progressive group *b* is contemporaneous with a conservative group more or less representing its structural ancestry, a common phenomenon in mammalian history and one of great helpfulness

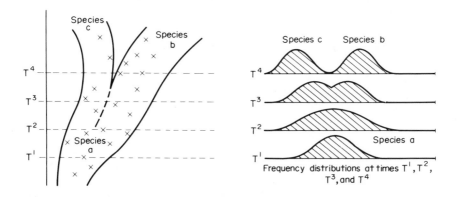

FIG. 4–7. The data of Fig. 4–3 interpreted as one species giving rise to two and the sort of theoretical frequency distributions expected at successive times if this interpretation is correct.

in reconstructing phylogeny if the true ancestral group has not been discovered. As drawn, the diagram more nearly approximates a pattern of this latter sort, *c* differing little from *a*, but the general principle would be the same if *c* were quite unlike either *a* or *b*.

The simplest types of successive frequency distribution patterns corresponding to such a history are also shown in the diagram. At an early period an adequate sample would tend to show, as regards many of its variates, a single and approximately normal curve, typical of a single phylum of average variability. Assuming in this simplified example that the splitting of the phylum takes place gradually and by the segregation into two stocks alleles which are likely to occur in any member of the parent stock, the next stage in time would tend to show frequency curves still single but exceptionally low and broad, technically platykurtic, corresponding to increased variability or mutability in the race. Still later, as segregation begins to be a visible factor, the curves will tend to be bimodal. Finally, the characters distinguishing the two descendent phyla will give two separate curves.

In every case of data of this general type, which is by far the commonest type with which a paleontologist has to deal in drawing up phylogenies, these three general sorts of interpretation are about equally probable *a priori*. Under the best circumstances and applied with proper judgment, the frequency distribution criteria here outlined in their simplest and most generalized form are unquestionably the most conclusive that can be employed. As yet, they are probably directly applicable only to a minority of our problems, although the available data are rapidly increasing. In default of these, there are other criteria nearly as useful, and still others of decreasing value until they approach guesswork, which also has a place in our literature. It would seem desirable to adopt the simplest interpretation, unless there are definite and strong reasons for adopting one more complex. A group may well be considered monophyletic until some contrary probability is established. This is in keeping with sound general principles of research and is also a salutary attitude in view of recent extreme trends in this particular field. As a point of less theoretical but great practical importance, it also tends to impede unnecessary complication of nomenclature and of taxonomy, which is the chief instrument of discussing phylogeny.

Source

(1937) Patterns of Phyletic Evolution, *Bulletin of the Geological Society of America*, Vol. 48, pp. 303–314.

Remarks on Vertebrate Phylogeny

Comments

In 1969 the First International Symposium on Animal Phylogeny (*I Simposio Internacional de Zoofilogenia*) was held at the University of Salamanca. That university, founded in AD 1230, is the oldest in Spain and among the oldest in the world still functioning. It never had been noted for general excellence in the sciences and for a time had fallen into a somewhat poor position, but now was reviving and hoped to spur the revival and to direct it in part of the sciences. It therefore invited and financed the attendance of zoologists from several different countries and entertained them royally. The speaking sessions were open and were well attended. I wrote my contribution in English and it was published in that version, but I read it in Spanish, which Spanish friends helped me to make more fully idiomatic. As it is a fairly clear expression of some ideas appropriate for the present section of this book, I include part of it here. This covers most of its topics, but I have deleted considerable parts, scattered throughout, because they were more appropriate for a largely local audience not very familiar with the subject or because they are sufficiently covered elsewhere in this book.

Text

The science of phylogeny requires the subjective reconstruction of objective but now unobservable phenomena. As a science, it must nevertheless be based on definite and repeatable observations. It has often been stressed that observations relevant to phylogeny are of two supposedly quite distinct kinds: those with and those without a time dimension, generally interpreted as meaning observations of fossils and of recent animals. Of course fossils are also observable only at the present time, but they are remains of animals which, when alive, occurred in a definite sequence, and in most cases the sequence has now been determined by geological methods within increasingly narrow limits of error. Nevertheless the study and interpretation of fossils often differ from those of recent animals only to the extent that some kinds of data on fossils are more restricted.

Fossils aid enormously in the reconstruction of phylogenies, but if they were indispensible, the aims of a general zoophylogeny would be chimerical. Darwin already pointed out with care and in some detail how comparative data on contemporaneous animals, and not only sequential data, can be interpreted in an evolutionary, including phylogenetic, way. That conclusion has only strengthened since Darwin and has acquired an even more extensive philosophical justification and methodological basis. Probably the most important recent developments regarding phylogenetic data for vertebrates are of nonclassical kinds that can only be obtained in contemporaneous animals and interpreted by comparative, not temporally sequential, methods.

Classical data for phylogenetic studies of vertebrates were almost entirely observations of gross or nonmicroscopic anatomy. Such data still are most important in this field, and present conceptions of vertebrate phylogeny derive primarily from anatomy. However, the most striking recent development is the increasing availability and use of information of many different sorts. It is not true that all possible observations are useful for phylogenetic studies, and the accumulation of data that are useful has a point of diminishing returns. Nevertheless the use of more kinds of data is generally helpful both in confirming and in correcting phylogenetic inferences earlier based almost wholly on gross anatomy. Nonanatomical evidence increasingly used includes: behavior, karyology, serology, protein sequences and other molecular data, and various other kinds.

If we had adequate samples of all the specific lineages and geographic races within a group of animals as fossils spaced at intervals shorter than the minimum time of morphological transition from one nominal species to another, then a true and complete phylogeny for that group would emerge automatically, objectively from a simple array of the data. There is no considerable group of animals for which such samples are known. In fact it is improbable that such complete fossil data exist for many or indeed for any groups as large as families. That is no argument against the factual nature of phylogeny. It is an inevitable result of the sampling processes of fossilization and recovery, never continuous or closely continual over extensive geographic areas and through extensive geologic times. That was already clear in a general way to Darwin and it has since been worked out in detail by paleontologists and is a basic fact of their profession.

Not even in the most favorable instances, such as the famous example of the horse family, do the data now known or, in all probability, ever to be known provide an absolutely complete, truly objective phylogeny at the level of species. That fact has two important concomitants, among others. First, when we reconstruct a phylogeny we are seldom doing so at best for a complete group of any considerable size and at the level of species throughout. For example, we do not know and are not likely ever to know the precise sequence (or, more probably plural sequences) of species by which reptiles gave rise to mammals. We do know that it was the reptilian Order Therapsida that gave rise to the Class Mammalia, and that is phylogenetic information of great interest and importance.

The second point to be made here is that even when there is an excellent fossil record, and of course to still greater degree in the more frequent cases where the fossil record is relatively poor or absent, phylogeny is not obviously inherent in the data. It must be inferred by a confrontation of the most and best available data with a body of interpretive principle and theory. There was a phase when many zoologists, and especially paleozoologists, believed that most phylogenetic problems might be solved by a set of rather simple rules or so-called "laws". For example, it was believed that larger animals do not give rise to smaller or complex animals to simpler. Evolution being considered irreversible, an animal with an enlarged tooth could not be ancestral to one with a smaller tooth. Evolution being considered orthogenetic, a phylogeny could not include a change of direction in an evolutionary trend. However, those and numerous similar rules are not laws but only statistical generalizations which cannot be assumed to hold good in any particular instance. It has indeed been found that some proposed "rules" have more exceptions than examples. Oddly enough, one of the extremely few rules that does not have exceptions was seldom stated and has been ignored in a few attempts to reconstruct phylogenies: a taxon occurring later in time cannot be ancestral to one earlier.

In spite of the fact that a search for a simple solution by rules was largely unsuccessful, a long series of combined theoretical and empirical researches has produced an acceptable working body of methodology for phylogentic inference. Other general statements have been given by taxonomists of the evolutionary school, which holds that taxonomy should be related to phylogeny.

Methodological and theoretical problems of phylogeny and of classification are closely related, even when an explicitly evolutionary taxonomy is not being sought. It is therefore often difficult but it is necessary to keep clearly in mind the fact that the two kinds of studies are not identical. A taxonomic key, for example, is preferably based on as few characters as possible and in most useful form it has no necessary or evident relationship to the phylogeny of the included animals. Even an evolutionary classification, constructed so as to be consistent with an inference as to most probable phylogeny, does not represent that phylogeny either in detail or in form. A tree figure, dendrogram, can be constructed so as to represent a great variety of relationships and not only those of phyletic descent. It resembles a phylogeny and is too often mistaken as such, sometimes even by its author, when it is not.

One source of confusion is that relationships represented often are those of single characters or single kinds of characters. Within limits, inferences as to phylogeny become worthy of increasing confidence in proportion as they are based on larger numbers of characters and more distinctly different kinds of characters. A methodological problem here is how to take so many different data into account simultaneously. For this purpose no instrument or method yet devised or now realistically imaginable even remotely approaches an experienced human brain. For instance simply by looking at, say, two fishes held next to each other a phylogeneticist simultaneously compares a number of dimensions and of morphological elements that is *literally infinite*. He has no trouble in taking into account at the same time many other kinds of data, on behavior, physiology, habitat, chromosomes, or whatever is available.

Here the question of quantification arises, including the quaint notion that nothing is really scientific until it is reduced to mathematics. That dictum is fallacious and can be carried to the point of being ridiculous. It has nevertheless been evident for centuries that some degree of quantification is essential in taxonomy, which has always included numerical data and methods. It was finally realized, at a comparatively late date, that because taxonomy must deal with populations as its subject matter, appropriate numerical methods are commonly statistical. In taxonomy, the methods that began to be adopted by zoologists, especially in and after the 1930s, were particularly useful for description, diagnosis, and identification. They made little contribution to phylogenetic or evolutionary aspects of taxonomy, which continued to be almost entirely nonquantitative. More recently and especially in the 1950s and 1960s, some zoologists began to adopt and also to devise special methods of quantifying resemblances among related taxa. These methods at first were not intended to assist in phylogenetic studies but on the contrary to eliminate them as far as taxonomy is concerned.

The usual procedure with those methods is to construct a dendrogram from which a nonevolutionary (strictly phenetic) classification can be devised mechanically. The dendrograms resemble and sometimes approximate two-dimensional phylogenies but cannot be taken as such.

Among the reasons why numerical phenetic dendrograms are misleading guides to phylogency are the following: they do not distinguish between homologues and convergent resemblances; they make no allowance for different rates of evolution of different characters and in different lineages; they frequently treat more characteristics of little phylogenetic significance (often easily quantified) while minimizing or omitting those of greater phylogenetic significance (often less obviously quantifiable); in the usual form, they reject weighting of characters that nevertheless do demonstrably have different weights (degrees of significance) in the light of evolutionary and phylogenetic principles. On the other hand, these methods instrumented by electronic computers facilitate the reduction of large masses of numerical data to forms not in themselves phylogenetic but usable for phylogenetic inference. That use is increasingly recognized and methods for it devised.

It remains true, and I suspect that it will always remain true, that all these methods and instruments are only aids to the one indispensible instrument for reconstruction of phylogeny: an instructed and experienced human brain. Although "subjective" has often been used as a pejorative epithet in this connection, all known methods have a large subjective element, overt or covert. The reliability or confidence level of the result depends more on the quality of its subjectivity than on the degree of its supposed objectivity.

Phylogeny and classification are distinct subjects, even for those who seek to base classification on phylogeny, but they are closely associated, even for those who seek to exclude phylogeny from classification. Taxonomic nomenclature provides an essential means of discussing phylogeny, and phylogenetic problems are generally definable in taxonomic terms. They also differ characteristically with taxonomic levels. A zoological taxon cannot be studied intelligently as such unless some attention is given to its probable phylogeny. The consideration of the phylogeny of species within a genus, for example, will differ greatly from that of orders within a class or classes within a phylum. There are literally thousands of studies of vertebrate phylogeny at lower taxonomic levels. It is impossible to review them and unnecessary to exemplify them in this brief general survey.

The most fundamental phylogentic problems, those at the level of phyla, are also among the most difficult and remain at present relatively low in confidence. Most of the basic transitions involved apparently occurred among small, soft-bodied animals few fossils of which have yet been discovered, although there is an excellent chance that more will be. Comparative studies of recent animals give rise to suggestive inferences, but in the case of the vertebrates, especially, we have only extremely modified representatives of the phylum, and the connections to be inferred are in an extremely remote past, certainly not less than 5×10^8 years ago and probably on the order of 10^9. It was inferred long ago and is still generally accepted that the tunicates (Urochorda) and lancelets (Cephalochorda) are more or less specialized offshoots from the ancestry of the vertebrates at a stage where distinctly vertebrate organization had not yet been achieved. A still less vertebrate-like group called Hemichorda or Stomochorda has often been considered an earlier offshoot of the same sequence. A theory now old held that all these groups, and hence ultimately the vertebrates, had a common ancestry with the earliest echinoderms. That theory has not been replaced by any other more probable, but little further evidence for it and some against it has appeared in recent years.

Although little useful study is now being made on prevertebrate phylogeny, new data on phylogeny at all levels among the vertebrates are accumulating at a rapid pace and debate on many aspects of the subject is lively and in general fruitful. Much recent discussion of phylogeny of the vertebrate classes has involved questions of monophyly and polyphyly. The question of monophyly and polyphyly has been greatly confused by failure to distinguish principles of phylogeny from those of evolutionary classification. It apparently cannot be repeated often enough that classification, even on the strictest evolutionary basis, *cannot express* phylogeny. A phyletic group, if correctly interpreted, has in principle been derived at some point from a single ancestral group, very likely a single species but not necessarily so: perhaps a deme or a subspecies, perhaps a species-group or even a genus-group. The relationship of that principle to a given classification, an intensely practical matter to those who do in fact practice it, is not always simple and is not really pertinent in this discussion of phylogeny except as it has confused such discussion.

A good example is provided by recent arguments about the Class Mammalia. Simpson, Van Valen, and Reed proposed quite different arrangements of the Classes Reptilia and

Mammalia, Crusafont later proposed another, and still others not necessary to this discussion have been advanced. Those mentioned were not based on any deeply significant difference as to phylogeny. All agreed that later mammals, by any definition, were derived from a large taxon usually called Therapsida and hitherto generally referred to the Class Reptilia. All agreed with earlier conclusions by Simpson, Olson, and a number of others that the therapsids included many phyletic lines which were becoming increasingly similar to later mammals and more than one of which independently acquired a structural–physiological level that had usually been taken, for practical taxonomy, as a convenient way to define mammals as distinct from reptiles. The transition was gradual, and any taxonomic line drawn between earlier and later groups is arbitrary. The difference of opinion about therapsid-mammalian classification is not phylogenetic but refers only to where it is believed possible and most desirable to draw such lines.

On the other hand, there is currently a difference of opinion on the same subject that also touches on taxonomy and is so expressed but that really does raise phylogenetic and not only taxonomic questions. In the middle and late Jurassic there are a number of known groups of vertebrates usually designated as orders of mammals. In common with most students, I would now recognize five: Multituberculata, Triconodonta, Symmetrodonta, Pantotheria, and Docodonta. They certainly had a common ancestry at some time. The truly phylogenetic and not only taxonomic questions are when that common ancestry occurred, at what evolutionary level, and, after that, in what sequence the divergence of the groups occurred. On much less evidence than is now available, I long ago suggested that separation into at least three lines occurred at the therapsid level and that only the symmetrodont–pantothere divergence occurred thereafter. For a time, later considerations suggested to me and others that there was an even greater degree of nominal "polyphyly", meaning not that they did not have a common ancestry but only that more lineages later advancing to mammalian structural level had become distinct earlier in the Triassic and among structural therapsids. Now I believe that the latest discoveries and studies change the situation. It is quite possible that all the nominal mammals known from the middle to late Jurassic, listed above in ordinal terms, arose from a single triconodont-like stock of the late Triassic with the exception of the Multituberculata.

I have stressed some points about early mammalian phylogeny at high taxonomic levels because I happen to be especially familiar with them and I find them highly instructive. There are equally interesting and even more hotly disputed phylogenetic problems involving the origins of other vertebrate classes. Virtually everyone now agrees that amphibians, or land vertebrates (tetrapods) in general arose from fishes, and from a particular broad group commonly designated as Crossopterygii. However, some students, mostly of the Stockholm school, believe that two phyletically quite different groups of fishes are included under that taxonomic term and that different amphibians, so-called, arose entirely independently from each. They believe the commonest living amphibians, the anurans (frogs, etc.) to be more clearly allied phylogenetically to reptiles than to other amphibians, the urodeles (salamanders, etc.) On the other hand, there is much evidence that anurans and urodeles are in fact related, that they had a common ancestry among structural tetrapods, and that the latter evolved but once, as far as known forms are concerned, at least.

I have referred to just two of the major problems, especially interesting just because they are not yet satisfactorily solved. Information on them continues to increase, and solutions doubtless will be reached. I need hardly mention that vertebrate phylogeny represents an enormous body of knowledge achieved, and yet also a fascinating series of things yet to be studied. Doubtless the most supremely interesting phylogeny is our own. Much there, too, remains hazy, but much is known.

Source

(1971) Status and Problems of Vertebrate Phylogeny, in ALVARADO, GADEA, and DE HARO, editors) *I Simposio Internacional de Zoofilogenía*, Facultad de Ciencias, Universidad de Salamanca, pp. 353–368.

Higher Categories in Phylogeny and Taxonomy

Comment

The expression "higher categories" is taxonomic. It refers to the hierarchic sequence long accepted as the framework for zoological and botanical classifications, from top to bottom or higher to lower: Kingdom, Phylum, Class, Order, Family, Genus, Species. (Complete classifications now usually have other categories between those of the classical framework.) The species is now generally held to be the basic unit of classification. Its nature and definition are discussed in a later section of this book. The higher categories are those above the species, progressively higher or more extensive from genus to kingdom. Groups of organisms defined and classified in the higher categories are higher or supraspecific taxa.

Although thus designated taxonomically problems about higher categories have involved also their phylogenies and beyond that the evolutionary explanations of their origin and deployment. A former debate over whether the evolutionary principles or factors involved in the origin of species, designated "microevolution," are the same as or different from those giving rise to higher taxa, "macroevolution," has subsided. While not quite universally it is now usually held that the processes are the same although they may be acting with greater or lesser intensity, over longer or shorter periods of time, and in different circumstances at different hierarchic levels.

The origin and nature of higher categories can be and have been discussed from several different points of view. The approach in the paper here reprinted in part is concerned primarily with adaptation and secondarily with classification. The original paper in full dealt with a number of groups as examples of taxa at various supraspecific levels. In the abridged version here presented only the following examples have been retained: the vertebrates as a whole, at the level of the classes of one phylum; the mammals at the level of a class derived from another; and the primates as one order within a class and as subdivided at levels from families to suborders. Some passages not essential for present purposes have been excised, but other changes from the original publication are minimal.

Although the text accompanying the present Fig. 4–8 in the original paper is not here reprinted, this figure is reproduced and now briefly discussed because it is methodologically interesting and might be more widely employed and because it has also a classificatory significance to which I paid too little attention at the time. This example involves six families all of which were for years, including the time when I wrote this paper, referred to the order Carnivora. All but the Felidae (cats) are extinct. In five of these families the cheek teeth evolved a shearing, carnassial, meat-cutting function by the development of a more or less sharp crest. One way to measure the degree of that carnassialization is the ratio between the posterior and the anterior edges of upper cheek teeth, premolars and molars. (In highly carnassialized dentitions the third upper molar is usually and both the second and third are sometimes lost.) The figure graphs these ratios for one genus in each family. Projecting above the base line measures the intensity of carnassialization and the shape of lines connecting the values for the separate teeth of one genus indicates the pattern of carnassialization.

In the figure the upper four examples all have strong carnassialization, the fifth has only slight carnassialization, and the bottom family has none at all. The patterns for

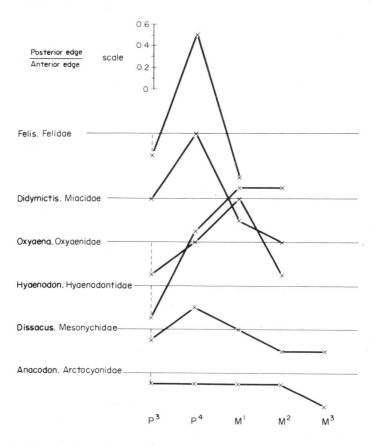

FIG. 4–8. Ratio of length of posterior edge to anterior edge of upper cheek teeth (third premolar to third molar as far as present) in one genus (and species) each of six families of mammals formerly referred to the order Carnivora. The scale applies to all six diagrams with the value 1.0 for each represented by the corresponding horizontal line. For further explanation see text.

Felidae and Miacidae are closely similar, and this is consonant with the conclusion that the Miacidae, broadly speaking, were ancestral to the Felidae. The next three patterns are different from each other and from the Miacidae or Felidae. They probably evolved carnassialization independently. It was of course known that the Arctocyonidae had not evolved carnassialization but it was formerly inferred that they conservatively retained a precarnassial stage in the origin of the Carnivora. It is now commonly held that they do not represent the ancestry of a monophyletic group Carnivora and are more nearly related to or belong among the primitive ungulates.

Text

Major natural or nonarbitrary higher taxa have typically arisen with (one could say "by") some important basic change in adaptation. The particular adaptive development that accompanies the origin of the taxon may later be overlaid and to some extent obscured by secondary adaptations and specializations. It may even be secondarily lost (e.g. in nonflying birds), but these complications do not invalidate the generalization. The nature and significance of the basic adaptation are extremely varied in different instances. Among many others, these broad special cases may be singled out for exemplification and special discussion:

The new adaptation may permit replacement of ancestral or other competing groups and then also frequently permits still wider exploitation and more specialized ecological subdivision of similar environments. Examples: jawed as against jawless fishes, teleost fishes against other osteichthyans, mammals as against reptiles.

The new adaptation may involve or permit entrance into distinctively different adaptive zones or ecological situations not available to ancestral forms. Examples: tetrapods as against fishes, birds as against reptiles, Pinnipedia (seals and walruses) as against Fissipeda (terrestrial carnivores).

In either of the preceding two cases, but particularly the former, the new adaptation may be a general improvement or complex new integration hardly analyzable in terms of single characters. That is true in the examples of the teleosts, the tetrapods, and the mammals. At lower levels there are also frequent examples, e.g. among the primates or in the hare family.

Again, in either case but particularly the latter, the new adaptation may involve a few distinct characteristics or even a single one, although this does usually also involve concomitant or subsequent broader adjustments and integration. That is true of the jawed fishes and the birds. Other examples are common, including the rodents and the artiodactyls. The Artiodactyla (an order of hoofed mammals), not otherwise discussed here, arose by essentially a single modification in the tarsus or ankle.

All of these factors are illustrated at a fairly high categorical level by the vertebrate classes (Fig. 4–9). The jawless fishes (Agnatha) were early almost entirely replaced by jawed fishes and survive only in the little-diversified, ecologically narrow and specialized recent cyclostomes. Details of the transition are almost undocumented, although its general nature is reasonably clear. Acquisition of jaws was accompanied by improvement in locomotion leading, among other things, to effective predation impossible to early agnaths. The first jawed fishes, placoderms, a diverse class probably polyphyletic in origin, were in turn rapidly replaced by higher or true fishes (chondrichthyans and

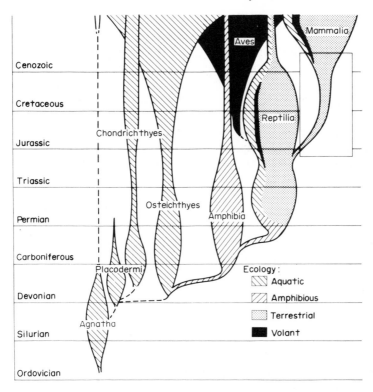

FIG. 4–9. Relationships and distribution in time of the classes of Vertebrata. Expansions and contractions of the various classes are not scaled but are suggestive of their diversities in the course of time. "Amphibious" ecology is used loosely for ecologies most usual in the Amphibia and some early reptiles. Some later reptiles and mammals are also amphibious in a rather different sense.

osteichthyans) with general adaptive improvement, especially evident in locomotion but undoubtedly also in general physiology. Invasions of entirely new environments, on land, occurred in a series of steps through early amphibians and reptiles. The most obvious change was evolution of legs from fins, which apparently occurred rapidly, but there were other equally important and, for the most part, slower changes. These are best documented in the skull, but there were also radical physiological changes such as the evolution of exclusive air-breathing. Facultative air-breathing was already present in the ancestral fishes. The amniotic egg, which could develop out of the water, had evolved by early Permian, at latest.

The birds reached another large series of new ecological, adaptive zones essentially by the evolution of a single new faculty, that of flight. The flying apparatus is, however, complex in itself, and concomitantly there were many associated adaptations: changes in the brain, acquisition of homeothermy, and others. These varied structural and functional changes occurred at different times and at different rates, a phenomenon that has been called "mosaic evolution" on the basis of this example. The term "mosaic" would be misleading if taken to imply quite independent evolution of different avian characteristics or lack of adaptive integration at any point in the process. Even *Ar-*

chaeopteryx, a comparatively clumsy animal with feeble powers of flight, was undoubtedly a well-integrated organism, but its integration was on a different basis from that of later birds. The full potentialities of feathered flight were not reached all at once but by the slow development of one system after another, each nevertheless remaining coadapted with the others.

Other reptiles, the pterosaurs, acquired flight in quite a different way and probably even earlier than the birds. Taxonomically they are considered an order of Reptilia rather than a separate class simply because their primitive reptilian heritage was not so completely transformed. They did not achieve the full potentialities of all the ecologies involving flight and they were vastly less diverse than the birds. Certainty is impossible, but it was probably competition with the more effective birds that caused the extinction of pterosaurs.

Evolution of mammals from reptiles was analogous with, for instance, that of osteichthyans from placoderms and not with that of birds from reptiles. There was no one key adaptation and no invasion of a wholly new adaptive zone. There was a whole series of general structural and physiological improvements that made mammals more efficient within some of the same ecological situations and that eventually permitted mammals almost completely to replace reptiles and also to subdivide the formerly reptilian zones still more finely and effectively.

A second and perhaps even more important point is that the general improvements, even though evolving in mosaic fashion, affected practically all the mammal-like reptiles, Therapsida, and early mammals. At the same time there were changes that were order-, family-, genus-, and species-specific. These were for the most part, at least, adaptive to increasingly narrow ecological situations and they diverged remarkably among animals that were at the same time evolving in parallel (at varying rates) as regards the more general improvements.

There is of course no sharp distinction between the kinds of adaptive change that I have called "general" and "specific." At one end of the scale are adaptations that will be advantageous for any animals that can (genetically) acquire them over a broad range of environments and ecological situations. At the other end are those adaptive only in one particular niche. Every intermediate exists, and these trends are correspondingly more widely or more narrowly distributed among larger and smaller taxonomic categories. A similar contrast of general and specific adaptations has already been exemplified at lower taxonomic levels among the rodents.

From the relationships between mammals and reptiles just described, it follows that the mammals are not a "natural" group as regards their origin. A horizontal division must be made at their base. Since no one position for this line will mark a crucial stage in all or most of the numerous general improvements, diagnostic of Mammalia, the division is distinctly more arbitrary than, for instance, that at the base of the Rodentia. Nevertheless in later times, notably at the present day, there is a wide structural and physiological gap between all mammals and all reptiles so that the taxa now are nonarbitrary. This is not seriously gainsaid by the fact that there is now a group (Monotremata) and were earlier several groups separated by large gaps from both mammals (in the strictest sense) and reptiles and classifiable, for practical purposes, as either or neither.

For practical purposes in arbitrary definition of the earliest mammals, few of the general improvements in physiology are determinable from fossils. The practical definition is based on the jaw and ear. In all recent reptiles the jaw is compound (has several different bones) and articulates with the skull by means of the articular and quadrate while the middle ear contains a single ossicle, the stapes. In all recent mammals the jaw is simple (dentary bone only) and articulates directly with the squamosal, while the middle ear contains three ossicles, the stapes plus the former articular and quadrate. Fossils show that even these now completely associated characters were acquired in succession, not all at the same time and not all at the same rates in different lineages.

Such general improvements crucial for the orgins of higher categories define what Julian Huxley has called grades, in contrast with clades, which are composed of a unified ancestry and its descendants and in taxonomic terms are monophyletic higher categories. Grades and clades may or may not coincide; or, in other words, the crossing of a grade line may be monophyletic (grade and clade coincide) or polyphyletic (they do not coincide).

The Theria, the present overwhelmingly predominant marsupial and placental mammals and their Mesozoic ancestry, crossed the reptile–mammal grade lines monophyletically at the ordinal level, at least. The mammals as a whole by any current definition arose polyphyletically in the sense used here, and just which forms are included in them depends on what particular detail is taken to define the grade. By contrast, the Rodentia are both a grade and monophyletic and that is probably (but not surely) true of the birds at the class level.

This situation raises taxonomic problems which make no particular difference to evolutionary interpretation of the data but are of considerable importance to systematists. With some doubtful details, the orders of Mesozoic mammals and the three basic later groups, monotremes, marsupials, and placentals, are nonarbitrary and well defined. As now known, each has its own characteristic basic adaptive and structural type and each is separated by structural–functional–ecological gaps from the others. Each is almost surely monophyletic. The class by all current definitions is nevertheless arbitrary, structurally and functionally definable only in terms of grades, and polyphyletic in detail. It would take too long to expound all the various possible taxonomic alternatives here. In accordance with my personal (but widely shared) bias for phylogenetic classification, I would however point out that classification based on phylogeny does not even by intention express phylogeny and is not even in principle rigidly monophyletic.

The Primates provide another example of the grade concept and its relationship to higher categories. The order Primates resembles the class Mammalia and the class Aves and the order Rodentia in that it arose by general adaptive improvement and not by any more or less clear-cut single basic adaptation. The most primitive Primates are distinguished only arbitrarily from primitive "Insectivora." In the usage intended, the primitive "Insectivora" are in fact primitive Cretaceous to Paleocene Placentalia or Eutheria that had not yet acquired the special characteristics of any more precisely definable placental order, including the Insectivora of later times. There is some tendency among recent students, mostly still *in schedis*, to confine the designation Insectivora to the later, comparatively specialized insectivores.

The broadest outlines of primate history can be drawn both in terms of radiations (Fig. 4–10) and of grades (Fig. 4–11). The basic radiation occurred mainly in the Paleocene, with various subradiations at later times. Although lineages resulting from the basic radiation were extremely diverse, and still are to a somewhat lesser extent, they may in a general way be taken as representative of a single, wide grade and labeled

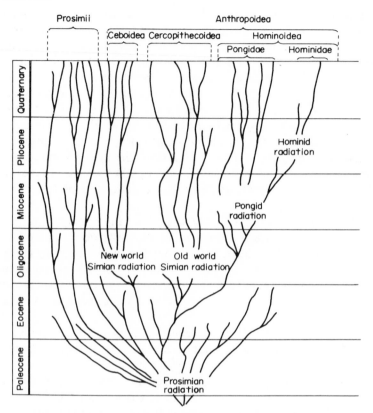

FIG. 4–10. Representation of a simplified phylogeny of the Primates as involving adaptive radiations at various times. The lines are schematic of clades but do not identify particular lineages or taxa. A suprageneric classification consistent with this phylogeny is given at the top.

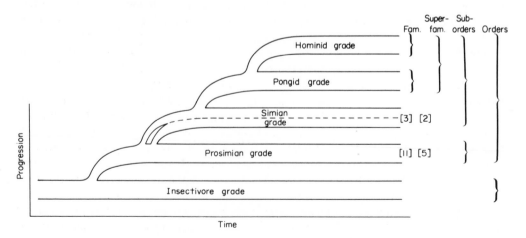

FIG. 4–11. Primate history schematically represented as a sequence of grades rather than as a proliferation of clades. Although there may be two or more phylogenetically distinct proliferations of clades within one grade, the grades are also reasonably consistent with a classification like that based on clades.

as prosimian. Recent representatives are the lemurs, galagos, tarsiers, and others.

Subsequent radiations, mainly in the Oligocene, probably occurred separately from prosimian ancestors in tropical America and in the Old World. Both reached about the same grade level, the existence of which is intuitively recognized in the vernacular term "monkeys." The grade may be called simian. It is evidently diphyletic, but both groups may have come from rather closely related prosimians. The simian grade is marked by rather conservative dental structure (less specialized than in many prosimians), reduction of the most primitive dental formula and comparative stabilization of the formula, some advance in brain structure, and various other characters.

Probably from one of the lines that separated near the base of the Old World simian radiation, another radiation occurred mainly in the Miocene of the Old World tropics. This involved increase in size, further development of the brain, loss of the tail, undoubtedly changes in the limbs although these are doubtful as to detail, somewhat specialized omnivorous teeth, arm-climbing (brachiating) locomotion in many members, and various other characters. The modern representatives are the gibbons and great apes, family Pongidae, and the grade may be called "pongid."

Finally, from some lineage that probably separated in the early part of the pongid radiation developed a poorly known and apparently quite restricted radiation, probably mainly in the Pliocene, characterized especially by upright posture with bipedal terrestrial locomotion and later by exaggerated development of the brain. So few lineages are known that this may not have been a radiation, strictly speaking, but rather only progressive advance in a rather unified group. The surviving product is *Homo*, and the grade is hominid.

The distinction of grades in this representation of the history is admittedly biased by human interest in *Homo*. Several grades of equal value with, say, the pongid grade could well be distinguished among the prosimians. Certainly as the grades here designated are followed upward they become much less diversified and also less widely distinct from adjacent grades. Those facts are reflected in most classifications by assigning the successive grades lower categorical rank. Thus, in the classification here followed, the prosimian grade is defined as one suborder and the three higher grades together as another. Five superfamilies are recognized in the prosimian grade, two in the simian grade, and only one in the pongid and hominid grades, together. The hominid and pongid grades are defined as one family each. The simian grade in contrast has three and the prosimian grade has eleven.

The evolution and classification of higher taxa have been discussed in terms of concrete examples at various levels.

These examples illustrate many different principles and generalizations, among which are the following:

(a) Higher categories generally arise by acquisition of a basic general adaptive complex, which may be retained essentially unchanged in all subsequent members (e.g. Rodentia) or may be profoundly modified or lost in some (e.g. Carnivora).

(b) Related but distinct higher categories at the same level generally arise allopatrically (e.g. later subfamilies of Leporidae) or allochronously (e.g. earlier and later subfamilies of Leporidae) or both (e.g. Ceboidea and Hominoidea among Primates). In any case they frequently later become synchronous and sympatric, with the result either of finer subdivision of ecological zones or of competitive extinction of one group and its replacement or relay by another (e.g. both outcomes among various reptiles *versus* mammals).

(c) The origin of a higher category may involve at one extreme (e.g. rodents, birds) a single character or adaptive complex or at the other extreme (e.g. primates, mammals) general adaptive change or improvement in numerous ways or in virtually the whole organization.

(d) In the former extreme (under (c)), subsequent evolution in the higher category often includes progressive coadaptation of related systems in one or more ways among most or all later members of the group (e.g. jaw muscles and lophiodonty in rodents).

(e) In either of the cases under (c), the final attainment of the fully developed characteristics of the higher category commonly proceeds at different rates and times in different systems or characters by "mosaic evolution," which must not, however, be taken to imply wholly independent evolution of different parts or lack of integration at any stage. The whole process is one of coadaptation. (E.g. first feathers, then gliding wings, then flying pectoral musculature, then loss of teeth, etc., in birds.)

(f) The same sort of basic or key adaptation may lead to different categorical levels in different circumstances. (E.g. gnawing incisors are ordinal in Lagomorpha and Rodentia but familial or lower in other orders: flight is a key class character in birds, but only ordinal in level among reptiles and mammals.) Determination of the level is a matter of judgment and consensus in which ultimate degrees of divergence and diversity, relative times of origin, and other factors (not here fully considered) are pertinent.

(g) The rise of a higher category may involve more effective exploitation of some or all of the adaptive zones of ancestors and collateral relatives (e.g. Mammalia among Vertebrata) or entrance into adaptive zones new for the group (e.g. Pinnipedia among Carnivora). In the former case, origin is usually by general improvement (see (c) above) and partial or complete replacement usually follows (b). In the latter case a key adaptive complex is usually involved and replacement, if it occurs, will be of more distant relatives.

(h) In the origin and evolution of many higher categories there is an interplay of more general adaptive improvements, often affecting many lineages in parallel, and of more specific adaptations of subgroups to particular ecological zones and niches. The general improvements may affect one (e.g. Therapsida) or even several (e.g. Mesozoic mammals) orders. The specific improvements may be, for instance, family-,genus-, or species-specific.

(i) The deployment of a higher category may be predominantly by general improvement, with broad adaptation and comparatively little diversification (e.g. Lagomorpha) or predominantly by specific and narrowing adaptation with much diversification (e.g. most Rodentia and Carnivora).

(j) Diversification in a higher category may be by divergent specific adaptations, often eventually becoming sympatric, or by parallel or convergent evolution of allopatric vicars (many examples of both in, e.g. rodents, carnivores, primates).

(k) The phylogenetic patterns in higher categories are diverse. The pattern may involve essentially a single lineage, simple successive splitting, or practically simultaneous multiple splitting, or any combination of these at different levels.

(l) Phylogenetic patterns may indicate a series of reasonably comparable natural or nonarbitrary taxa at a given categorical level or may be so diverse that no such series exists at the given level. Some orders do (e.g. Carnivora) and some do not (e.g. Rodentia) have natural suborders.

(m) Within a higher category, evolution often (but not always) proceeds by a sequence of adaptive radiations, which may provide a basis for natural subcategories at different levels depending on the scope of the radiation.

(n) Radiation commonly follows achievement of a new grade of organization, and in any case grades tend to correspond with higher categories. They may be reached monophyletically or polyphyletically.

Source

(1959) The Nature and Origin of Supraspecific Taxa, *Cold Spring Harbor Symposia on Quantitative Biology*, No. 24, pp. 225–271.

Supraspecific Variation and Higher Categories

Comment

In 1936 six specialist societies held a joint symposial session in Atlantic City under the broader egis of the American Association for the Advancement of Science. By invitation of the American Society of Naturalists I contributed to the symposium by presenting a long talk that was in effect two distinct although related papers on supraspecific variation and higher categories. The first part, here reprinted almost in full, was devoted to general principles, hypotheses, and conclusions drawn from a particular example among fossil mammals. A shorter excerpt from the second part of my contribution is given in the next portion of this book.

On the whole I believe that the discussion in the present text has stood up quite well. Further discoveries and studies have of course made some of the details outdated. For example, more notoungulates are now known, their anatomy has been studied in more detail, and their classification is somewhat changed. Nevertheless I believe the conclusions here drawn to be still basically valid. Some of the expressions, such as the "accretion of mutations," now seem vague or a bit naive, but not decidedly incorrect. A more sophisticated genetical approach to such studies was incipient and was later made.

Two points in particular made here have been subject to discussion at later dates, usually without reference to this paper, and that does make this reprinting still timely. One is that organisms as a whole or the various characteristics of organisms have not evolved at equal or at constant rates. That is true really beyond serious doubt on the evidence of the fossil record as regards the bodily, anatomical or phenetic aspects of organisms. Yet some biochemists and molecular biologists maintain that molecular evolution in DNA and proteins has proceeded with "remarkable constancy" of rates. I suspect that they are wrong, or perhaps wrong for some molecules and right for others. Even if they should be right for some, I maintain that molecules evolving at constant rates cannot have dominated phenetic evolution and therefore are of minor interest for explaining the phenomena of organismal evolution.

The other special point is that "groups comparable in taxonomic scope or in structural distinction or variety did not necessarily arise at comparable times," and that hierarchic level of supraspecific (or for that matter specific) taxa cannot logically and practically be based on time of origin or length of existence. I briefly mentioned supporting examples. Yet thirteen years later and again in 1966, Hennig, a German entomologist, published a book insisting on just those criteria of rank in classification. He cited the present paper but evidently did not believe it. Since that revised English version of his book his ideas have had some earnest, occasionally even fanatic, supporters. That particular idea is absurd, and a thousand more examples of its absurdity could be given. I add just one more: a little group of living sea shells (brachiopods) has been in existence since the Silurian at least but is ranked by all competent authorities as a genus, *Lingula*. It is older than several classes and should therefore by Hennig's criteria rank as a class all by itself. But it is obviously, closely, and monophyletically related to other taxa which by those criteria would be variously ranked as anything from separate classes down to separate species.

Text

The nature of paleontological evidence bearing on supraspecific variation and the general method of this approach to that complex problem can most easily be exemplified by selecting one group of extinct animals and mentioning a few aspects of variation in that group. Following a suggestion by Professor Gregory, I shall outline some observations and opinions involved in an intensive study of the extinct mammalian order Notoungulata.

The Notoungulata were a peculiar and varied group of herbivorous mammals, most of them hoofed, almost entirely confined to South America. They were abundant there throughout the Tertiary Period, but became rare and finally extinct during the Pleistocene. Their remains were first discovered by Darwin during the voyage of the *Beagle*, over a century ago, and the first descriptions were by Sir Richard Owen. By 1890, along with the usual crop of synonyms and names based on practically unrecognizable fragments, about ten really valid and well-characterized genera were known, all of later Tertiary or Pleistocene age.

The first genus recognized, in a sense the type of the whole Order, was *Toxodon* Owen, 1837, from the Pleistocene. It was a large animal, about the size of a small rhinoceros and of similar stocky build. Its incisors were specialized into an apparatus suitable for cropping vegetation, and its grinding teeth had very high, strongly curved crowns, whence the name ("bow-tooth"). Another classic genus *Nesodon* Owen, 1847, is manifestly related to *Toxodon* but is older, Miocene, and is smaller, more slender and in general more primitive. Both genera belong to one of the usually recognized primary divisions of the Notoungulata, the Suborder Toxodonta.

Another group of classic genera, including *Typotherium* Bravard, 1857, *Interatherium* Moreno, 1882, and *Hegetotherium* Ameghino, 1887, is based on smaller animals with gnawing anterior teeth and high, prismatic grinding teeth. These false-rodents are related to the rhinoceros-like *Toxodon*, but because of their rodent-like aspect and a few more detailed characters they have been united in another suborder, Typotheria.

Homalodotherium Flower, 1873, from the Miocene, is a large animal comparable to

Toxodon in size but with extraordinarily different proportions, its clumsy legs elongated in a peculiar way and its toes ending in large claws, rather than hoofs. Its teeth are low-crowned, in continuous series and relatively unspecialized. *Homalodotherium* is certainly a notoungulate, but it is so unlike the toxodonts or typotheres that it is generally placed in a third suborder, Entelonychia.

In the Toxodonta and Typotheria the limb structure is almost the same throughout, with rather unimportant modifications mostly correlated with bulk and speed. Their type of limb structure is primitive for the Notoungulata as a whole and in them never underwent any profound modification. Their teeth, on the other hand, became highly specialized, and in this respect the later genera are almost unrecognizably different from the earlier. This radical change occurred rapidly, between the early Eocene and some time in the Oligocene, then becoming stabilized and not undergoing any very important further changes before the extinction of each group, some of them in the Pleistocene.

In the Entelonychia the teeth never became so highly specialized. There were changes, of course, but in essentials the latest entelonychians, in the Miocene, had dentitions of early Eocene type. The limb structure, however, was deeply modified, and there is reason to believe that the major changes in this respect were rapid and concentrated in one part of the history of the group. Successive genera that are almost identical in the dentition are profoundly different in the limbs, just as some successive genera of typotheres are hardly distinguishable in the limbs but very distinct in the dentition. (See Figs. 4–12, 4–13.)

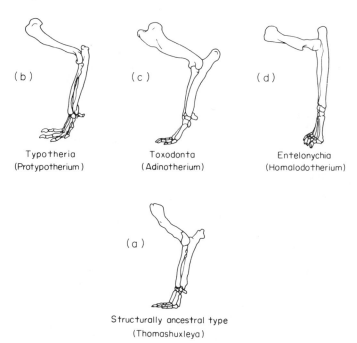

(b) Typotheria (Protypotherium)

(c) Toxodonta (Adinotherium)

(d) Entelonychia (Homalodotherium)

(a) Structurally ancestral type (Thomashuxleya)

FIG. 4–12. Lateral views of the forelimbs of four notoungulates, brought to equal size for comparison. The early Miocene forms shown were contemporary. The typothere and toxodont limbs were relatively conservative, but *Homalodotherium* (although now also classified in the Toxodonta) had become rapidly specialized in the limbs.

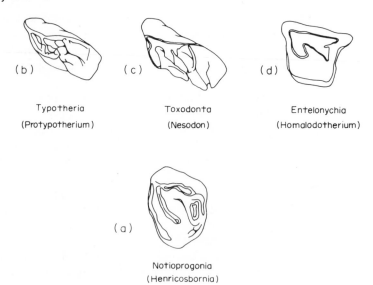

(b)	(c)	(d)
Typotheria	Toxodonta	Entelonychia
(Protypotherium)	(Nesodon)	(Homalodotherium)

(a)

Notioprogonia
(Henricosbornia)

FIG. 4–13. Crown views of left upper molars of four notoungulates brought to equal size for comparison. *Henricosbornia* represents a basic primitive pattern for the Notoungulata. The molars of *Homalodotherium* were conservative, but those of its contemporaries *Protypotherium* and *Nesodon* were already specialized.

These examples illustrate phenomena of wide importance in the study of supraspecific variation. Within a single group, such as the typotheres, one set of structures, such as the limbs, may be relatively invariable, evolving slowly and remaining primitive, while another set, such as the teeth, in the same group may be highly varied and variable, evolving rapidly and becoming specialized. Two different supraspecific groups, such as the toxodonts and the homalodotheres, may be essentially distinguished not so much by the general conservatism of one or by its uniform divergence, as by the fact that different parts of the organism remain primitive and different parts are specialized in the two groups. Moreover, degree of variability and concomitant rate of change in a progressive set of structures, as in the typothere dentition, is not necessarily constant but frequently characterizes a limited period and is preceded and followed by long periods of relatively low variability and little change.

Students of phylogeny often tacitly or explicity assume that a constant rate of evolution within one group or within related groups has been the rule, but if so it is a rule with so many exceptions that it is not a safe guide. Differences in rate of evolution between different structures in the same phylum, between different periods in the history of one phylum, and between different phyla with a common origin have certainly been very common. These different rates of structural modification, which are generally accompanied by different degrees of variability, are among the most important factors in the differentiation of supraspecific groups.

From the nature of these differences in rate, it follows that groups comparable in taxonomic scope or in structural distinction or variety did not necessarily arise at comparable times. One family of mammals, Didelphidae, arose in the Cretaceous and another, Ursidae, in the Miocene, an enormous difference, and yet by other and more useful criteria than time of origin these are both properly ranked as families in the taxonomic hierarchy. Such facts invalidate some inferences or prejudices regarding mammalian history, such as the argument that the family Hominidae must necessarily be comparable in antiquity with the other families of primates or that it must have developed at a comparable rate or at one equally constant.

These considerations lead farther to the taxonomic problem as to whether different groups of the same formal rank, for instance families, are or can be made equivalent by some criterion such as time of origin. It would take too long a digression to discuss this problem here, but it may be affirmed that this time criterion, at least, is impractical and quickly leads to confusion and to absurdity. Absolute equivalence between families (or other units) of different zoological divisions, such as fishes and mammals or, *a fortiori*, insects and mammals, does not exist and is probably quite unattainable, but a reasonably satisfactory working compromise can be reached.

Returning to the Notoungulata, the three-fold division into Toxodonta, Typotheria and Entelonychia, which has been outlined above, arose about 1895 and in various forms this general, simple arrangement has been widely used ever since. It is an admirable system as far as the well-differentiated and relatively impoverished faunas of the Miocene to Pleistocene alone are considered. Even in the Miocene the characters of the various phyletic lines within these groups were definitely fixed and the problem of the origins and of the basic relationships of groups hardly exists as long as attention is confined to these relics.

As early as 1897, however, the great Argentine paleontologist Ameghino began the description of much earlier faunas, some from well down in the Eocene. Many of these earlier South American notoungulates do not fit into the three usual suborders, and they give a very different picture of the development of the Notoungulata from that based on the later forms. Ameghino's solutions of the problems presented have not been accepted in their entirety by other workers. It would be unfair to attempt a résumé of his views here, for they were complex and full justice would demand a somewhat detailed analysis of the many largely unappreciated but really valuable and enduring parts of his work, and the distinction of these from the more superficial but more obvious points on which he is now generally conceded to have erred. The point most pertinent here is that in almost every group of notoungulates Ameghino observed two sorts of morphological resemblances: one to other South American groups of earlier or later age and the other to various Holarctic groups. The observation was, in general, both accurate and acute and all the resemblances mentioned exist and are important. In interpreting these results he concluded that both resemblances were evidence of actual phyletic relationships, so that the various Holarctic ungulates were considered as derived from analogous notoungulates. This theory, or rather complex of theories, has had to be abandoned, for it has since been demonstrated that the resemblances to Holarctic groups are almost entirely superficial and adaptive and not due to any close relationship.

An instructive example is provided by the Notohippidae, a family of Notoungulata which, as the name implies ("southern horses"), was supposed by Ameghino to be related to, and to be the source of, the Equidae of our hemisphere. The resemblance is largely confined to the dentition and is most noticeable in the development by both groups of high-crowned, prismatic cheek teeth, enveloped in a heavy layer of cement, and with a complexly folded pattern of enamel crests which is mechanically or functionally very similar in the two groups. Now that the history of these teeth is very well known for the Equidae and can be inferred with a high degree of probability for the Notohippidae, it is shown beyond any question that the development is independent but closely parallel in the two groups. (See Fig. 4–14.)

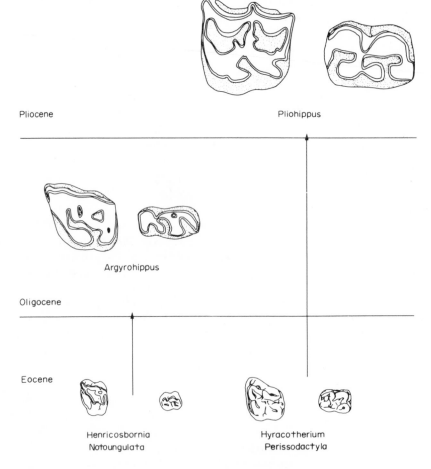

Fig. 4–14. Functional convergence of molar teeth in some notoungulates, left, and horses, right. Crown views of left upper and right lower molars, drawn to scale. In the early Eocene the ancestral types of both groups had small, low-crowned, cuspidate, cementless, browsing molars. In both groups some later forms evolved larger, high-crowned, crested rather than cuspidate, heavily cemented (stippled tissue), grazing molars. This happened millions of years earlier, and much more rapidly, in these notoungulates than in any horses.

Both start with low-crowned, cementless, sublophiodont teeth in the early Eocene. These initial stages, like the final development, are functionally closely analogous. There are, however, differences in the exact relationships and structures of the individual elements, cusps, crests and valleys, of the teeth in the two families, and these distinctions are maintained throughout their history. On this heritage, which is relatively invariable within each group, there are imposed new, progressive characters, mainly hypsodonty, lophiodonty and a cement investiture, and these are the same in both groups. It seems that there is a basic and relatively constant structure within each line to which are added new factors, or an accretion of mutations, that are the same for both but that appear in them independently. This independence is further emphasized by the time and rate of appearance of these mutations. The true horses continued with relatively little change in these respects through the Oligocene, underwent a pronounced acceleration in the

Miocene, and reached essentially the terminal condition at about the beginning of the Pliocene. The notohippids underwent a rapid development, perhaps at about the Miocene rate of the Equidae but at a much earlier time, in the Eocene. In the Oligocene they had already reached a condition about equal, in these respects, to that of the horses in the Pliocene. The notohippids then became extinct, despite the development of what would seem to have been the most efficient herbivorous dentition in the world at that time. Without straying too far into the tempting bypaths of speculation, may it not be that this remarkably early and rapid dental development was not matched by the correlative progress of other necessary factors and that the organism as a whole was thrown into a fatally unbalanced condition?

Aside from a rejection of Ameghino's views, which in many cases was unfairly sweeping, little constructive work on the early notoungulates was done by any one else until recently. Now the whole subject is reopened on a new and much more satisfactory basis. Ameghino's priceless and classic specimens, which were never definitively published and which a combination of circumstances long made unavailable for research, have now all been restudied in the light of later knowledge. Important new collections have been made, greatly increasing the number of specimens, permitting a good study of variation for the first time, revealing skulls and even skeletons of animals hitherto known only from isolated fragments and unearthing a number of new species and larger groups and at least one fauna older than any described by Ameghino. These and other factors have produced a revolution of ideas concerning this great mammalian order. In the state of flux into which the subject is thus plunged at present, conceptions of supraspecific variation and principles of phylogeny and taxonomy, which tend to be taken for granted during the more gradual evolution of knowledge, are brought into prominence and subjected to a severe test on a grand scale.

One of the first conclusions to emerge from this recent research was that many early notoungulates, indeed the majority in the oldest Tertiary formations, do not enter in a natural way into any of the three classic divisions of the Notoungulata: Toxodonta, Typotheria, Entelonychia. There had been some tendency to refer these to the Entelonychia, because they had been known almost entirely from the teeth, which resembled those of later entelonychians more than they did those of later toxodonts and typotheres. Now it plainly appears that the earlier toxodonts and typotheres also had teeth of this same general type, or in other words that supposed entelonychian characters in the dentition were merely primitive characters of all notoungulates, which happened to be less modified in entelonychians than in most other groups, as has already been suggested.

It appears rather that there was toward the beginning of the Tertiary a somewhat unified but extremely varied group, a sort of reservoir of phylogenetic potentialities, from which the classic groups derive. Within this ancient notoungulate complex there were numerous different minor phyletic lines, each of which was tending to be modified in its own direction and at its own rate. The classic groups merely represent such few of these lines as were successful in surviving into the Middle and Late Tertiary. Concordant with their longer history, these groups eventually came to differ more markedly from each other and from the common ancestry than did the much more numerous lines within the Early Tertiary complex. As a convenient means of designating these manifold, relatively primitive earlier forms I have proposed that they be classified as constituting a suborder of notoungulates under the name Notioprogonia. The notioprogonians include the structural ancestry or basic stock of notoungulates generally, and they also include a number of divergent phyla which died out before either their structural distinctions or their morphological variety were sufficiently great to warrant their being set aside in special suborders in a practical scheme of taxonomy.

An outstanding characteristic of the Notioprogonia is extraordinary variability, manifested in two ways difficult to distinguish but apparently really different: individual or infraspecific and phyletic or supraspecific.

In some cases it has now been possible to bring together large collections all of one geologic age and from one locality. With such material it is possible to recognize and delimit natural minimum group units, conventionally recognized as species, with considerable assurance and exactness. It is impossible here to go into details as to the criteria and methods employed, especially as they are rather laborious and many of them are new to paleontology, but I believe that biologists generally will recognize their validity. The general purpose and result was to recognize a sample of a population in which many variations appear, as in all populations, but in which these variations were arising in a single interbreeding community and with little segregation in definite components of the population.

An example is afforded by a sample of last upper molars of *Henricosbornia lophodonta* Ameghino, a notioprogonian common in the Eocene Casamayor Formation. A good, unified sample of 33 specimens shows a range in width of 5.8 to 7.1 mm, a difference of about 20 per cent, with an approximately normal distribution around a mean of 6.41 ± 0.06. Other dimensions vary to the same degree, and the variation in structure or in nonnumerical characters is also well marked, as is suggested by the accompanying figure [Fig. 4–15] of selected teeth showing the most abundant (or modal) structural type and also the relatively rare extremes of deviation from this. Different as these are, they all intergrade, and it is also noteworthy that to some extent the differences can be analyzed into single factors, such as round or long metaconule, attached or detached metaloph, strong or weak crochet, etc. In general, these show no significant association with each other, that is, groupings made on one of these characters do not tend to include the same individuals as those based on others. This is one of the lines of evidence that the sample is unified and does not consist of two or more distinct segregated zoological phyla. Differences in the other parts of the dentition are analogous in kind and in degree to those shown.

A single student, Ameghino, with all the material then known before him, based

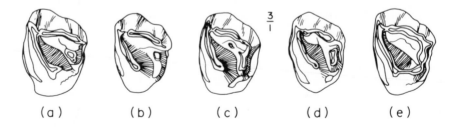

(a)   (b)   (c)   (d)   (e)

FIG. 4–15. Left upper third (last) molars of a primitive notoungulate. Although these all belong to a single species, *Henricosbornia lophodonta*, they have variations similar to those that later came to characterize distinct genera or taxa of still higher categories.

three families, seven genera and seventeen species on specimens all of which now seem almost certainly to belong to this single species, *Henricosbornia lophodonta*. The case is extreme, perhaps unparalleled in taxonomic history, but such multiplication of synonyms, on a smaller scale, occurs in the work of every active paleontologist and is in some degree unavoidable in this science. The statement is no indictment of Ameghino's ability, which was exceptionally great, but it introduces two important considerations: the need for and difficulty of obtaining an idea of the extent of variation within any one paleontological species, a topic outside the scope of the present paper, and the fact that the difference between infraspecific and supraspecific variation is often, or essentially, distributional rather than either qualitative or quantitative.

As regards this second point, Ameghino was using criteria which were valid in the groups from which they were derived and the application of which in this case, where they are not valid, followed what is still considered sound practice and was opposed by no apparent *a priori* consideration. The fact is that the differences between several of these variants of *Henricosbornia lophodonta* are exactly analogous both in kind and in degree to differences which do sometimes distinguish groups of generic and even of family rank, that is, they may be distinctive and relatively constant characters of different phyla and groups of phyla, each including several species, although here all are mingled in one species. For instance, so simple a character as the degree of attachment of the metaloph to the ectoloph is relatively constant in the Equidae, and differences in this respect in that family may define successive genera or may differentiate contemporaneous phyla, yet in *Henricosbornia lophodonta* every condition from almost complete independence (Fig. 4–15B) to almost complete union (Fig. 4–15C) occurs. Moreover, in notoungulates differences analogous to these may also characterize quite distinct supergeneric groups. For instance, in the Notohippidae two groups of genera, one typified by *Rhynchippus* and one by *Notohippus* itself, are differentiated by a relatively constant difference in the height on the crown at which two minor enamel folds, first and second cristae, merge.

The difference between the morphological characters of a small unit, called a species for present purposes, and of a group category, genus, family, order, including several of these units, may thus be quite different from the usual picture (which is, however, also valid in other cases) of the larger group being constituted simply by accretion of successive individual variations or mutations. In the basic unit, each character varies within the whole group as such. Every character has, so to speak, a certain repertory of variability which involves all the individuals of the group. This is true of the variations shown in *Henricosbornia lophodonta*, and these differences are infraspecific in this case. The higher categories are defined by characters which still vary, inevitably, but the variation of which is not confined to such narrow limits, is not from one individual to the next but is relatively constant within each small unit and shows its range of variability from one species to another.

From this point of view, *Henricosbornia lophodonta* is a potential ancestor of several different genera which would at first differ from each other little more than do the individuals of this species. If one or a few of these variations were segregated into one line of descent, so that variations of this particular sort always occurred, or more frequently occurred, in that particular line, a supraspecific evolutionary phylum diverging from the parent group in a fixed direction would arise. Continued elevation of variations from the individual to the group category would lead to continued divergence.

Source

(1937) Superspecific Variation in Nature and in Classification from the Viewpoint of Paleontology, *American Naturalist*, Vol. 71, pp. 236–267. (The pages here reprinted are 236–249.)

The Reality of Higher Categories

Comment

Few people now realize that Alfred C. Kinsey was an entomologist specializing on gall wasps of the genus *Cynips* before he reached a wider audience by his studies of human sexual activities. While he was still a systematic zoologist he studied more than 35,000 specimens of insects in this single genus and an estimated 126,000 of their distinctive galls. One of the voluminous results of that Herculean effort was a 334 page tome on the origin of higher categories, which generalized his views derived from *Cynips*. His broad conclusion was that although "species are realities in nature . . . all of the higher categories are artificial conventions . . . hardly real either in manner of origin or in their intrinsic qualities." He then listed by number the thirteen concepts of higher categories which he believed to be then current although badly outdated, and he maintained that his studies disproved all of them.

For the 1936 symposium on higher categories it was suggested that I discuss Kinsey's thirteen points, which I did one by one in the second half of the paper of which the first half has been reprinted above. My conclusion was that although some of the thirteen concepts were unsound, those were not really current and that in general Kinsey's treatment had not clarified the situation or contributed markedly to better concepts of higher categories. Discussion of Kinsey's views was dropped soon thereafter, and perhaps that contributed to his radical change in field and methods of study. In any case there would not now be much point in reprinting my treatment of all the thirteen points, but I do here give the discussion of one point, the sixth of Kinsey's thirteen, which then seemed and still seems to me the most important.

Text

"The higher categories are realities in nature. They were once real species." This is the most important point involved in the present discussion. Kinsey strongly opposed this principle and most of its implications. Most paleontologists will agree with Banks in believing that the limits of taxonomic categories are often undefinable in practice, since in the nature of evolution one grades into the other, but that both species and higher categories are real.

To demonstrate the conviction that supraspecific groups do have an objective reality, I have taken the liberty of copying a figure given by Kinsey (here Fig. 4–16) to prove the opposite contention, leaving the objective data as he gives them but changing the taxonomic interpretation. The dichotomous development with which the paleontologist deals has dichotomy in time, whereas that shown by Kinsey is developed in space, but

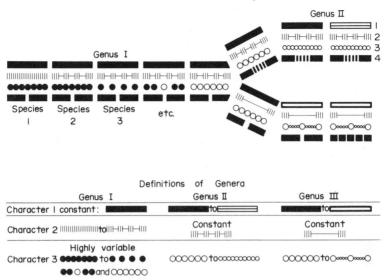

FIG. 4–16. The relationship of higher categories to the characters varying between species. The basic diagram, above, is the same as one published by Kinsey (1936, Fig. 14) as evidence for his contention that genera (and supraspecific taxa in general) are not real. The definitions and interpretation here given in terms of the characters in this diagram support the opposite view. For further discussion see text.

the difference is not essential. This dichotomy is an objective fact, assuming that the data have been correctly interpreted, and is exactly the process by which new groups are supposed to have arisen from one ancestral species. The status given these groups, species, genera, families, or whatever it may be, is arbitrary but the groups themselves are natural. Kinsey's figure might be interpreted in any of several ways, taxonomically, but that here given seems to establish the point. Kinsey's own interpretation is not in accordance with the reality of higher categories, but it seems open to objection to make such an interpretation and then to argue from it that categories are unnatural, when a natural arrangement of the data can, in fact, be made.

In this instance there are generic characters that are constant within each group designated by me as a genus. Thus the combination of characters shown as 1 and 4 is diagnostic of Genus I, 2 and 4 are diagnostic of Genus II, and 2 is diagnostic of Genus III. It happens that Kinsey has shown these as constant within the given groups. If this were not true, it would be more difficult to define the genus, but the reality of the genus would not be changed, nor would definition be impossible. For instance, Genus II can be defined not only by the constant characters 2 and 4, but also by the changes in 1 and 3 which take place in this line and in no other. The figure also shows the point that in a given stage, for instance, the first species of Genera II and III, the characters common to the divergent species, such characters here being 1 and 3, occur in the one species ancestral to both of them. To the best of my belief, however, no paleontologist or zoologist would really support the view, which is stated by Kinsey only for the sake of demolishing it, that these characters 1 and 3 must then appear unchanged in all the descendants of that species.

The paleontologist is familiar, more than the zoologist, with the usual absence of discontinuities in phylogeny, and he does not suppose his taxonomy to rest on a basis of discontinuity. What it does rest on in theory and in practice as far as the latter is successful is, however, a sort of discontinuity, that between branches, as may be seen in Kinsey's diagram. They are separate, and their union both to a single base does not mean that the separation is artificial.

In reference to this figure one more point must be made, which is that such a figure can be derived from actual data only by selecting a few characters deliberately such that they all (four in this case) differ in the end. There will be vastly more characters in which the terminal species do not differ appreciably, characters which are, in Kinsey's example, common to all *Cynips*, those common to all wasps, those common to all insects, and so on, for of course such characters also exist and vastly outnumber those distinctive of the species (See Fig. 4–17). They are perhaps to be taken for granted, but they are of the nature of the problem, for they do probably represent a general inheritance from a common ancestry and they are relatively invariable characters defining objective supraspecific categories.

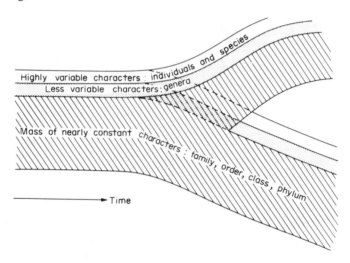

Fig. 4–17. One way of looking at intra- and supraspecific variation in evolving and dichotomizing (speciating) lineages. The principal point here involved is that when species of common ancestry separate and develop their distinctive characteristics, they still have many more characters in common than not. The characters that remain in common are generally ancestral and may tend to continue without much deviation as the group evolves into taxa at successively higher hierarchic categories. The streamlike diagram is a generalized conceptualization and must not be taken in too rigid a sense. Given characters do not necessarily remain in one of the formalized channels of this diagram. They are constantly passing from what are here conceptualized as the channels of lesser or greater variability, and that is a phenomenon often involved in the origin of higher categories.

Source

(1937) Superspecific Variation in Nature and in Classification from the Viewpoint of Paleontology, *American Naturalist*, Vol. 71, pp. 236–267. (The excerpt here is from pp. 255–259.)

Another Relevant Reference

KINSEY, A.C. (1936) *The Origin of Higher Categories in* Cynips, Indiana University Publications, Science Series No. 4, pp. 1–334.

The Concept of Species

Comment

Every thoughtful systematist has had some personal concept of the species as the basic category in classification and indeed in almost any discussion of organisms. Many systematists have expounded their concepts in the enormous, often repetitive and sometimes blatantly redundant literature of the subject. Although I have added to that literature more than once, the paper here reprinted with slight modification has been my most extensive discussion of the subject. It also makes a point that can be here somewhat clarified in retrospect. The so-called biological or genetical definition of the category species, although useful and valid in many applications, is essentially non- or even pre-evolutionary. In this paper I proposed a truly evolutionary definition of the species as a phyletic lineage evolving independently of others. I also pointed out that the genetical species may be understood as a special case of the evolutionary species.

In the latter part of this paper I further considered the special problems of species concepts in paleontological taxonomy. The analysis there given, especially in Fig. 4–19 and its discussion in the text, has implications that turned out to be wider than here appear. That analysis showed that a criterion for species almost simultaneously proposed by the German entomologist Hennig is impractical and illogical not only in paleontology but also in neontology.

This paper was stimulated by the flurries of discussion and controversy mentioned in the first paragraph. It is here reprinted essentially in full but with some deletion of references to now long-past controversial publications.

Text

The species concept is focal in evolutionary studies and, indeed, in all biological thought. Its endless discussion is sometimes boring and seemingly fruitless, but is not wholly futile. In the course of such discussion the concept has been clarified, comprehension and a consensus have tended to develop, and the concept has changed in a significant way. There have recently been two more flurries of attention to this perennial topic. One was originally concerned with whether the species is a "fiction" or is "objective," but also treated such matters as the relationships of neontological and paleontological species concepts. The other was concerned with the bases and practices of paleontological systematics and also with "morphological" versus "phylogenetic" or "natural" versus "unnatural" classification.

I believe that it is possible to combine some of the apparently conflicting views into one consistent statement, and to add a few significant considerations not explicit, at least, in the recent discussions.

Parts of the discussions and some of the apparent conflicts are primarily semantic. By a ponderous application of symbolic logic, Gregg sought to show that an issue raised by Burma and Mayr is not a genuine taxonomic problem or, at least, that if it does relate to a taxonomic problem it does so in the wrong words. It is, of course, important that words be used as accurately as possible and that they do not obscure properly taxonomic questions. Nevertheless, Burma and Mayr (as well as some other discussants) *were* considering a genuine taxonomic problem, in words perhaps not logically impeccable but, taken in context, adequately performing their main semantic function, that of communicating understandably among colleagues.

The semantics of the systematists' vocabulary is a fascinating subject, which surely does have its own importance but which has the danger of merely diverting attention from the systematists' proper business, systematics. I believe that most of the purely semantic confusion on the present subject can be avoided if such terms as "real," "natural," or "objective," and opposite or contrasting terms, are not applied to taxonomic categories or methods of classifications, and if the two terms "arbitrary" and "nonarbitrary" are used in specially defined senses.

Definitions of taxonomic categories, such as a species, specify the sort of data or of inferences from data that are to be used in assignment of organisms to a group ranked in that category. For instance, the category definition of a species as a group of "actually or potentially interbreeding natural populations which are reproductively isolated from other such groups" (Mayr) specifies that data and inferences as to interbreeding and its absence are to be used. In some cases the data or inferences used will indicate essential continuity among the organisms to be grouped, and in other cases they will indicate essential discontinuity. Under the preceding genetic definition, actual or potential interbreeding is continuity and reproductive isolation is discontinuity. With a morphological–associational definition, continuity would be overlap in variation between compared populations and discontinuity the absence of overlap. Essential continuity or discontinuity in georgraphic, ecological, or temporal distribution has obvious meaning.

I propose to call taxonomic procedure arbitrary when organisms are placed in separate groups although the information about them indicates essential continuity in respects pertinent to the definition being discussed, or when they are placed in a single group although essential discontinuity is indicated. Conversely, procedure is nonarbitrary when organisms are grouped together on the basis of pertinent, essential continuity and separated on the basis of pertinent, essential discontinuity. Of course no stigma is meant to attach to taxonomic procedure thus defined as arbitrary. A completely nonarbitrary classification is impossible. It would be possible to extend discussion to such points as the precise definition of "essential continuity" or other parts of these definitions, but I think their meaning will be clear to all taxonomists, now or as discussion proceeds, and that semantics may be dropped at this point.

The typological concept of a taxonomic group is that the group corresponds with an abstract or ideal morphological pattern. Variation may be dealt with by a fixed or intuitive standard as to allowable deviation from the pattern, in which case the grouping is arbitrary. (It may either include discontinuities or draw a line across continuity.) Or, somewhat less naïvely, at a given level, usually that of species, the criterion of continuity in variation around the pattern may be used, a nonarbitrary procedure for that category.

The typological concept is pre-evolutionary and non-evolutionary. It still underlies a great deal of taxonomic practice but is now seldom favored in theory. The only serious modern theoretical support for frankly typological taxonomy comes from those few students who believe that species arise by abrupt morphological change from one "morphotype" to another.

Most of the data actually used in the practice of taxonomy are morphological. It is therefore not surprising that practical taxonomists suggest from time to time that classification should be morphological, in principle, but a purely morphological classification would be based strictly on degrees of morphological difference between organisms, and this is really so impractical that no one really tries to do it consistently. It is a commonplace that the degree of morphological difference within what everyone, morphologist, geneticist, or other, calls a single species is frequently greater than that between what all call separate, related species. It is also quite impractical to obtain a valid, overall measure of total morphological difference between two organisms. Characters are always selected, weighted, and interpreted. Even typological classification, more strictly morphological than others, requires definition of the morphotype from characters in a group already set up on grounds not, in practice, purely morphological. Typological or not, practical morphological classification starts with some sort of grouping and in most practice this is usually an attempt to recognize what is (whether so called or not) a genetically defined population.

The fundamental point here for taxonomy is the modern idea that it is populations, not specimens, that are being classified. I have insistently recommended population concepts in taxonomy, and so have many others. Some paleontologists have been rather slow to grasp or accept the population concept of taxonomic groups. Despite some conservatives and reactionaries, the concept is already widely accepted among neontologists.

If classification is to start with populations, category definitions at and below the species level should refer to populations which, further, should be meaningful biologically. It seems to me, and few systematists are likely now to question this, that such groups should likewise have evolutionary significance. Here is the most serious fault of typological or of purely morphological definitions. Unless by chance or unless a hidden genetical criterion is actually used, they do not define biological populations or have clear evolutionary significance.

Attention to biologically significant populations is the basis and justification for the now usual neontological definitions of the species category in terms of interbreeding and reproductive isolation, i.e. of genetical factors, like the definition already quoted from Mayr. The fact that a species, as a group, is actually diagnosed in morphological terms, does not conflict with definition of the species, as a category, in genetical terms. The basis for definition of a category is quite different from the evidence available for decision as to whether a particular group of organisms meets that definition. And although the evidence used is mainly morphological in practice, it also almost always includes other sorts of data as well: distribution or association, at least, and preferably also other information.

The genetical definition of a species as a group of actually or potentially interbreeding organisms reproductively isolated from other such groups is nonarbitrary both in its inclusion and its exclusion. Its criteria are reproductive continuity and discontinuity. The group defined is coextensive with the continuity and bounded by the discontinuity. A species under this definition is the largest group with nonarbitrary exclusion and the smallest group with nonarbitrary inclusion. By the criteria of this definition and in cases to which it applies, infraspecific groups are nonarbitrary as to what they include (being reproductively continuous, by definition), but more or less arbitrary as to what they exclude (having boundaries without full reproductive discontinuity, by definition). Under the same criteria and circumstances, supraspecific groups are arbitrary as to inclusion, because by definition they do or may include two or more groups between which there is discontinuity, but nonarbitrary as to exclusion, because their boundaries coincide with the nonarbitrary boundaries of included species.

Thus under this particular concept and in the particular cases to which it applies, the species is defined as the one taxonomic category that is nonarbitrary both in exclusion and in inclusion. This is another way of expressing what is clearly intended by statements that the species (so defined) is the "objective" or the "real" taxonomic unit. If my usage of "nonarbitrary" is accepted and discussion of the meaning of "objective" or "real" is avoided, it should not be seriously questioned that the statement of the first sentence of this paragraph is valid. Objections, which may also be entirely valid in their own terms, are of five principal sorts:·

(1) The genetical concept of species is not the only one possible, and for certain groups and in particular circumstances it may be less desirable than some other.

(2) Application of the genetical definition to actual cases, even those to which it could theoretically apply, sometimes turns out to be vague or impractical.

(3) There are many groups of organisms, or circumstances involved in their taxonomic grouping, to which the stated genetical definition does not apply even in theory.

(4) The genetical definition implies but does not adequately state or overtly take into consideration more definitely evolutionary criteria on which it does or should depend, criteria as to the evolutionary role of a lineage, to be discussed below.

(5) Application of this or of related evolutionary concepts of the species does not correspond with past and current usage in certain groups and by certain taxonomists.

It seems to me that all these objections have considerable force, more force than is granted them by some students whose taxonomic work is in the circumscribed fields where the genetical definition is in fact most practical or those whose interests are not primarily taxonomic. Yet I do not think that the objections invalidate the genetical concept or remove it from a central and basic position in taxonomic theory. They merely require that it be modified in certain applications and that it be supplemented by other concepts to meet situations to which it does not properly or practically apply. The rest of this paper is devoted mainly to discussion of some desirable or necessary modifications and supplemental concepts.

As Mayr (1946, 1950) has emphasized, the usefulness of the genetical concept in taxonomy and the nonarbitrary definition of the genetical species (its "objective reality") are most evident in what he calls "nondimensional species," those established in biotas living in one place at one time. Under such conditions, discontinuities in morphological and associated physiological variation are usually evident. In sexually reproducing groups it is almost always easy under these circumstances to establish by observation, experimentation, and inference which morphological discontinuities reflect reproductive discontinuities and to designate these as species boundaries.

But, as Mayr has also recognized, the fact that genetical species are usually rather obvious under these special limitations does not mean that they are equally clear and the genetical definition equally adequate under other and perhaps more important conditions. Populations do have extension in time and space and a nondimensional taxonomy cannot cope with many essentials of life and of its evolution. With extension in space, the criteria of genetical continuity and discontinuity, of actual or potential interbreeding or its absence, cease in many cases to be absolute and clearly nonarbitrary and become merely relative. The similar and related local populations may not in fact interbreed over a period of years and yet may reasonably be considered as still having that potentiality. On the other hand, quite extensive interbreeding may occur between adjacent populations which nevertheless retain their own individualities, morphologically and genetically, so clearly that any consensus of modern systematists would call them different species. In some groups of plants, even though species are defined and considered as genetical entities, occurrence of some hybridization between adjacent species may be the rule rather than the exception. In such cases the species are in part arbitrarily bounded even though the gene flow is less between than within species, and the genus may become the most fully nonarbitrary unit. A rigidly genetical zoologist might then insist that such botanical genera equal zoological species, but evidently most botanists feel that in some way their species *are* analogous with zoological species and they can make out a good, even though not an absolutely clear-cut, case.

In practice, even by zoologists who adhere strictly to genetical concepts of taxonomic units and who work on groups to which the concepts are applicable, it is often clear that the criterion of interbreeding or its absence is not taken as wholly decisive. Species may be distinguished even though they interbreed (hybridize) to some extent, and populations may be referred to a single species even though there is evidence that they are not in fact interbreeding. Other criteria are given weight additional to that of their evidence on interbreeding, e.g. morphological divergence, partial or full intersterility, and especially occurrence with discontinuity in the same area.

The genetical definition is meaningful because it is related to the evolutionary processes that give rise to the groups being classified. Yet the genetical criteria are not related to evolutionary change directly but only, as a rule, by implication. The following seems to be the strictly *evolutionary* criterion implied: a phyletic lineage (ancestral–descendent sequence of interbreeding populations) evolving independently of others, with its own separate and unitary evolutionary role and tendencies, is a basic unit in evolution. The genetical definition tends to equate the species with such an evolutionary unit. Most of the vagueness and differences of opinion involved in use of the genetical definitions are clarified, at least, if not wholly resolved by taking the genetical criterion, or interbreeding, not as definitive in itself but as evidence on whether the evolutionary definition is fulfilled. Thus the species as actually used by many progressive systematists in both animals and plants does tend to approximate a unitary phyletic lineage of separate evolutionary role even though in both cases outbreeding, hybridization, may occur and in some groups of plants this is widespread and usual. Emphasis on unitary evolutionary role also essentially resolves the theoretical difficulty of defining species in asexually reproducing groups.

This redefinition, or shift of emphasis, or revealing of the implicit basis of much modern evolutionary taxonomy, introduces the element of time into the concept of species, even in the so-called nondimensional situation. It designates the species, including the "nondimensional" species, as a unit which has been evolving separately, or which will do so, or, as a rule, both. Decision that populations will evolve separately involves prediction. Such points as wide geographical discontinuity (especially with a strong intervening barrier), morphological divergence, sympatric occurrence without interbreeding, and intersterility are clearly items of evidence for this sort of prediction. Their bearing seems to me more meaningful in evolutionary terms than in the definition of actually or potentially interbreeding populations, although of course the evolutionary species usually is also such a group. The special importance of intersterility, even though no modern taxonomist makes it an absolute requirement for specific separation, is, for instance, evident in this context: intersterility makes the prediction of separate evolutionary roles certain.

Part of the endless discussion on species concepts is concerned with the relationship between neontological and paleontological species. Opinions vary from the view that the two usually are quite different to the view that they are usually essentially the same or that one is only an extension of the other concept. Both views are correct in the sense that species just like those (by any definition) of neontology do occur in paleontology, but that actual practice regarding them may be more difficult or, at least, necessarily somewhat different in paleontology and that there also occur in paleontology taxonomic groups to which no strictly neontological species concept can properly be applied.

That paleontological data and materials are different from neontological is well known and sometimes overstressed. Direct genetical methods are unavailable in paleontology, but they are very rarely used in neontological taxonomy. The paleontologist usually has parts, only, of the organisms concerned, but the neontologist commonly uses parts, only, of recent organisms. Different parts may be available or used in the two cases, but inferences from them regarding populations may nevertheless be closely analogous or actually identical. Nearly or exactly the same general sorts of data, morphological, distributional, and associational, are frequently used in the practice of paleontological and neontological taxonomy.

The "nondimensional" species is encountered more frequently in paleontology than in neontology, in spite of the fact that paleontology is inherently more multidimensional than neontology. The neontologist is seldom forced to confine himself to collections from one locality, and is never justified in doing so unless forced. Much paleontological taxonomy is necessarily and properly based on quarry collections or mass collections from one local stratum, associations without appreciable dispersion in space or time and ideally nondimensional. In such cases neontological concepts and definitions of genetical and evolutionary species apply without modification.

Discontinuities are more frequent and of more varied sorts in paleontology than in neontology. This has certain disadvantages for paleontological theory and interpretation, but it also has some practical advantages. Discontinuities of observation, only, due to inadequate sampling of local populations or inadequate distribution of sampling stations, occur in both fields but are generally harder to fill in when paleontological. Discontinuities of record, that is, in the organisms actually present and available for sampling in the field, are a particular paleontological problem and may concern both time and space. When samples have been obtained from different localities or horizons, rocks and fossils intermediate between them may not exist. Such discontinuities are, as of now, facts in nature. Their use to delimit taxonomic groups is nonarbitrary, by definition. Yet they do not necessarily coincide with any particular sort of discontinuity that existed when the organisms were alive. Hence their relationship to the sorts of units defined in neontology may be and remain ambiguous. A special case is illustrated in Fig. 4–18.

The special questions involved in succession or sequence will be discussed separately, but it should be noted here that paleontological samples discontinuous in space are often also discontinuous in time and that the possibility can seldom be discarded. With such samples of similar organisms it is always difficult and it may be quite impossible to determine whether:

(a) They represent local populations that were genetically continuous, and hence infraspecific groups by genetical and evolutionary definition.

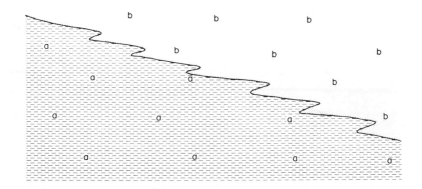

FIG. 4–18. An example of complication and apparent falsification of the fossil record. Broken lines and dots represent two sedimentary depositional facies with related fossils, *a* and *b*, each confined to one of them. In any single exposure *a* is always below *b* and hence will be considered older, possibly ancestral, but in fact *a* and *b* are contemporaneous.

(b) They represent separate phyletic lineages, and hence distinct genetical and evolutionary species.

(c) They represent ancestral and descendent populations, and hence a special and peculiarly paleontological situation discussed below.

In such a case, the preferred practical procedure is:

(1) To consider the two (or more) lots of associated specimens as samples of different local populations and to derive from them estimates of morphological variation in those populations.

(2) If the population estimates indicate no significant mean difference, to consider the samples as representative of essentially a single population and hence taxonomic group.

(3) If the population estimates indicate significant mean difference but overlap in range of variation, to consider the samples as drawn from different subspecies of one species.

(4) If the population estimates indicate no overlap in range of variation (for at least one well-defined character), to consider the samples as drawn from different species.

Species recognized in this way are nonarbitrary in exclusion and inclusion by combined morphological and distributional criteria. They are morphologically similar to most genetical and evolutionary species. In many cases they will in fact be genetical or evolutionary species, but under the stated conditions it is virtually impossible to determine this equivalence with any high degree of probability.

Succession on a small scale and involving short periods of time occurs in some neontological data and there involves some special taxonomic problems, but on the whole succession is distinctively paleontological.

Discontinuities of observation and of record are frequent in paleontological study of successive populations. They frequently correspond with discontinuities of time, already mentioned. Diastems, geologically brief intervals of nondeposition (with or without erosion), are abundant in most stratigraphic sequences. They represent local discontinuities in time, but may be considered taxonomically unimportant if there was no significant change in the populations being studied or if intervening fossils of the same or closely similar populations are available from other localities. Larger and regional stratigraphic unconformities are also, although less, common and they usually represent taxonomically significant discontinuities in time.

Discontinuities in succession may also be caused by migration, by change of (biotic, and commonly of correlated stratigraphic) facies, or frequently, by a combination of both. Such discontinuities often coincide with discontinuities in time but they need not do so.

When discontinuities in succession are present in the data, they may be dealt with in practice as outlined above for paleontological discontinuities in general. They may similarly permit nonarbitrary delimitation of morphological–distributional species which may approximate, but cannot usually be clearly equated with, genetical–evolutionary species. This greatly simplifies paleontological procedures and in many particular cases it averts the special taxonomic problems inherent in continuity of succession.

Essential continuity in sequences long enough to involve significant progression or diversification of populations is far from universal in paleontology. It is, however, frequent and becomes steadily more so as collecting becomes more extensive. The special problems involved therefore do have great and increasing practical importance. They are of supreme importance for paleontological taxonomic theory. No one seriously doubts that the whole of life has factually been a continuum of populations when the whole sequence is considered, in spite of the innumerable discontinuities in the record.

The genetical–evolutionary concept of species is applicable as between different phyletic branches, evolving lineages, especially if they are contemporary but also if they are not. Thus in Fig. 4–19A, *a*, *b*, and *c* are three different species, by genetical or, more clearly, by evolutionary definitions, although *a* and *b* are contemporary with each other but not with *c*. Serious problems in theory, and usually also in practice, arise rather regarding the parts of such a pattern that cannot be distinguished as separate branches.

One possible solution, diagrammed in Fig. 4–19B, is to recognize central lines as species and to distinguish branches as other species. This procedure is "correct" from an evolutionary point of view, or, better, the species so designated do fulfill the proposed evolutionary definition even though their delimitation is genetically arbitrary at the points of branching. For rather small groups under exceptionally favorable circumstances the procedure is also practicable and is actually used. Its practicability depends, however, on recognition of an essentially unchanging central line, *a*, and main branches, *b* and *c*. It is, however, more usual even within rather small groups and universal within really large groups and long sequences for all lines to evolve materially. Then it is not practical taxonomy to designate the whole of any one line as a single species, and there is no meaningful criterion for designating "main" or "central" lines and branches. Thus the four alternatives of Fig. 4–19C are all equally acceptable interpretations of the same phyletic facts as in Fig. 4–19A, in terms of main lines and branches, if all lines are undergoing progressive change. The only reasonable criterion of choice would be designation of certain terminal branches as more important, or somehow definitive, than others. A logical extreme would be, for instance, to take *Homo sapiens* as the supreme species and to consider its ancestry, from the beginning of life (or even before) as the main line, not specifically separable from *H. sapiens*. This arrangement has in fact been

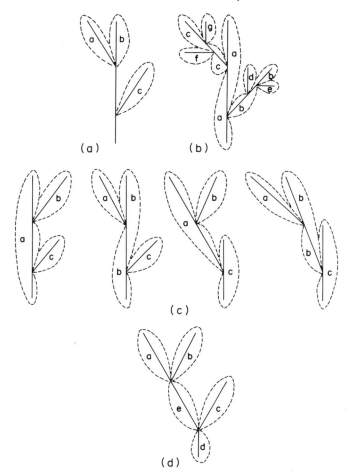

FIG. 4–19. Classification of populations in branching phylogenies. Time sequence is from bottom upward and outward. The solid lines represent lineages of descent and the broken lines enclose segments as they might be classified as specific taxa. A has three species by almost any definition but leaves parts of the phylogenetic sequence unclassified. B is entirely divided into species with a persistent central species, *a*, two also but less persistent successive branches from it, *b* and *c*, and terminal lesser branches from them, *d*, *e*, *f*, and *g*. C is topologically identical with A but is completely classified in four alternative ways which depend on which lineal succession is considered central or found to be more persistent. D is also topologically identical with A and C but is divided into species using branching as the only criterion, which requires definite knowledge of all branches. See further discussion in the text.

seriously proposed by a philosopher. Taxonomists will surely agree that this result and the whole procedure involved are impractical if not absurd.

Another possible approach is to recognize each evolutionary lineage as a unitary species until it divides and then to consider the descendent branches as species distinct from each other and from the single ancestral line, as diagrammed in Fig. 4–19D. This grouping meets an evolutionary definition of species, although delimitations between adjacent species are arbitrary by genetical criteria. It is, however, both undesirable and impractical. It frequently happens that a population undergoes no essential change even though a branch, a separate species, has arisen from a part of it. E.g. in Fig. 4–19D, *d* and *e* may be genetically and morphologically identical in all essentials. It is then not meaningful taxonomy to designate them as separate species. An even more serious objection is practical: the pattern of branching in a paleontological sequence is gradually discovered, perhaps never fully known, and generally depends as much on opinion as on unequivocal data. The taxonomy of long-known species would be changed every time a new branch was discovered or inferred and would be excessively and unnecessarily subject to personal disagreement. Moreover, a phyletic line may change radically between branches (say within *e* of Fig. 4–19D), and it is then not useful taxonomy to classify it as the same thing throughout.

The difficulties involved here are merely obscured by the presence of phyletic branching. They arise, regardless of whether or not branching occurs, from the problem of classifying ancestral and descendent stages in a continuously evolving population. Such a population may be diagrammatically represented, as in Fig. 4–20A, by a curve of variation (both genetical and morphological), moving through time and also being displaced as its genetical and morphological characters change. A cross-section represents the population at a particular instant in time, as it would be represented by a fossil

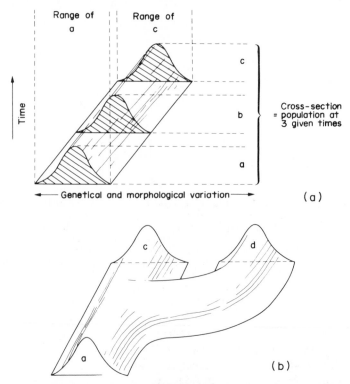

FIG. 4–20. Classification of successive populations. The varying and successively changing populations are symbolized by normal curves at given times in A and by the solid generated by moving normal curves in B. A, sequence without branching, leading to a population at *c* with variation not overlapping that of its ancestor at *a*. B, branching sequence in which one lineage runs from *a* to *c* as in A and a branching lineage leading to a separate species, *d*, contemporaneous with *c*. See further discussion in the text.

sample from a single horizon. Such a cross-section is a genetical nondimensional species, as seen both in neontology and in paleontology.

The whole sequence of populations, *a* to *c* in Fig. 4–20A, is genetically continuous and it fulfills the conditions of both genetical and evolutionary definitions of a species, as previously discussed. By these concepts, it is a single taxonomic group, defined as a species. Yet with the passage of time and continuation of progressive evolution, *c* has become quite different from *a*. For purposes of evolutionary study and of practical application to stratigraphy, it is essential that a distinction be made between these populations, which are different.

In practice, the paleontologist calls *a* and *c* different species if, as in Fig. 4–20A, the inferred ranges of variation do not overlap. They are *not* different species by the widely accepted genetical criteria or by the proposed evolutionary criteria discussed above. The comparison is clarified in Fig. 4–20B in which speciation (in the neontological sense) is represented as also having occurred. In this diagram, *c* and *d* are different species by any current usage; explicitly they are different genetical–evolutionary species and also different species in current paleontological practice. But *a* and *c* are parts of a single genetical–evolutionary species, although called different species in paleontological practice.

The paleontologist thus uses the designation "species" for two sorts of entities which are radically and fundamentally incongruent. The only way in which the species category might be defined so as apparently to include both sorts of entities would be to abandon any evolutionary significance for taxonomy and to use purely morphological criteria. But this is not a useful solution. The general undesirability and impracticality of *purely* morphological taxonomic concepts have been sufficiently emphasized above. Moreover, the *whole* sequence of organisms represented in Fig. 4–20B cannot be classified at all, in morphological or any other terms, if the pattern in time, i.e. the evolutionary situation, is ignored. As static, separate pictures, the morphological difference between *a* and *c* and that between *c* and *d* are of the same sort, but within the pattern of the whole group in time, even the morphological relationships are not the same in the two cases, for *a* and *c* are morphologically (as well as genetically) continuous through intervening populations and *c* and *d* are not.

In the situation represented in Fig. 4–20A, the desirable and indeed necessary taxonomic separation of *a* and *c*, whether they are called species or by some other category term, is arbitrary, because through intervening stages they are continuous by all meaningful criteria. The placing of an intermediate population, such as *b*, in one category or the other is, of course, also arbitrary. When the data really reflect the continuity of the sequence, intermediate populations must often be placed by rule of thumb rather than by any more positive and meaningful criterion. It is in such situations that the frequent occurrence of discontinuities of record, the absence of part of the sequence *a–c*, is practically useful in providing a means of separating *a* and *c*. This is still a separation in what *was* a continuum, but it is nonarbitrary (by the special definition of that word in this paper) as regards the actually available materials being classified.

Since paleontologists are applying the designation "species" to two fundamentally dissimilar sorts of taxonomic categories, it would appear logical that they confine that name to one of them and use a different name for the other. This has also been suggested, but it runs up against another serious practical difficulty: the paleontologist often does not know and has no way to determine which of the two basically different sorts of groups called "species" he has before him.

It is a common situation to have two discontinuous paleontological samples such as *a* and *c* of Fig. 4–21A. (It has been noted that if *a* and *c* are discontinuous in space, the possibility that they are also different in time can seldom be ruled out.) By applying the practical methods previously summarized, the paleontologist can readily draw population inferences from these samples, find that variation probably did not overlap in the populations, and define them as different "species." However, he does not know in what sense they are different species, because he does not know whether the relationship is as in Fig. 4–21B or as in Fig. 4–21C, and unless other crucial populations can be sampled he may have no conclusive way of finding out. In dealing with different samples from closely similar populations, this is one of the commonest situations in the practice of paleontology.

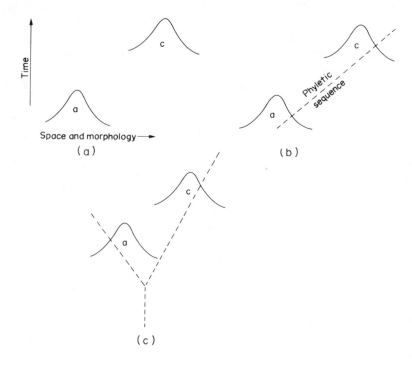

FIG. 4–21. Different possible interpretations of two paleontological samples separated in space and (or, and possibly) in time, as shown diagrammatically in A. B, interpretation as a single lineage. C, interpretation as two specific branches from a common ancestry. See further discussion in the text.

In such cases, a distinction cannot be made in practice between "species" in the basic genetical or evolutionary sense and in the sense of subdivisions in a continuous ancestral–descendent line. I do not here favor or propose a special term for the latter sort of taxonomic group. I do maintain that it is desirable and useful to realize that these are two quite different things, and that the "species" of paleontological taxonomy may be of either sort.

Source

(1951) The Species Concept, *Evolution*, Vol. 5, pp. 285–298.

Other references

MAYR, E. (1942) *Systematics and the Origin of Species*, Columbia University Press, New York.
MAYR, E. (1949) The Species Concept: Semantics versus Semantics, *Evolution*, Vol. 3, pp. 371–372.
SIMPSON, G.G. (1961) *Principles of Animal Taxonomy*, Columbia University Press, New York.

Types, Name-bearers, and Hypodigms in Taxonomy

Comment

Having achieved some concept of the term "species" and other categories in the hierarchy of classification, the systematist who is naming new species or revising old ones has to cope with a routine which is set by custom and by official codes of nomenclature. There are several different codes, but that relevant for zoologists and paleozoologists is now the International Code of Zoological Nomenclature adopted by the XV International Congress of Zoology, cited below under Stoll *et al.* (1964). Anyone engaged in these pursuits soon runs up against the ambiguity of the term "type," an ambiguity that is not removed, is indeed additionally confused, by the Code.

The essay here reprinted in part was an attempt to end that ambiguity, which I believe that it did in the most sensible possible way. These usages and the term and concept of "hypodigm" here first introduced have become widely but not yet quite universally accepted by zoologists, especially paleozoologists. The Code refers to the "type" as "the type of a taxon," whereas I and many other zoologists maintain that a type, in the sense of the Code itself, is the type of a name (or nomen) and not of a taxon. In the terms here used, it is a name-bearer and not a standard of reference. For some time now several members of the International Commission of Zoological Nomenclature, which administers and regulates the Code, have attempted to get the Commission to adopt this clarification but have been thwarted by other commissioners. Systematists can also be conservative or, not to mince words, pigheaded.

This is not merely a legalistic quibble. As the following extract also indicates, it involves clarity of thought, of expression, and of concepts and procedures in classification. It further involves a philosophical point as to just what a species is and an untangling of three decidedly different concepts all of which have borne the label "type."

In this reprinting I have omitted brief introductory remarks not necessary for present purposes and also the last few pages of the original essay. Much of that part of the essay discussed the terminology related to types. A dreadful maze of more than fifty terms for different sorts of "types" had grown up, and most of them were and are worse than useless. I remarked that some of them had been proposed "more for their prestige value than for a more scientific purpose" and that "'Every specimen a type' is the curatorial equivalent of 'every man a king.'" The height of absurdity had been reached with "icotype," a specimen about which nothing had been published.

The Code requires the use of the terms "holotype," "syntype," "lectotype," and "neotype" under particular circumstances. There is neither need nor cause for other terms for types considered solely as name-bearers, as they should be.

Apart from the omissions at beginning and end, I have here changed a few words. In the original essay I several times used the word "theory" in a way I now consider ambiguous, and I have replaced it by "principle" or by "view."

Text

There are three basic taxonomic operations in which "types," in the broadest and, I think, most confused sense of the word, are concerned. For the moment attention may be confined to types that are actual specimens, individual animals or parts of them, and to the taxonomic categories, species and subspecies, based on these. A species, as it is actually defined or diagnosed and used in the literature is a subjective concept. The usual view, often questioned but held by me and by most taxonomists, is that this concept corresponds more or less with a real thing in nature, a group of individual animals that are truly related in a way that makes them a natural unit of a certain approximate scope. This thing in nature may be considered the real species, but it is not and cannot be the species of taxonomy. The mental concept that is the species of taxonomy cannot be shown to be coextensive with the real species and even if by chance the two were coextensive, it would be an error to suppose them identical. One, the taxonomic species, is an estimate of the other, the real species.

Although it is a subjective concept and not a material thing, this estimate is not drawn out of the void but is based on observations of concrete specimens. These specimens do not constitute the real species but they are selected on the basis of belief (whether well- or ill-founded) that they do belong to one natural specific unit. The description of them is not a description of a species unless it implies or expresses a belief that they represent a natural group of which they are only a part, usually a relatively small part. The species concept is poorly formed, or indeed really absent, unless an attempt is made to infer from the specimens the probable characters of other animals that belong to the group but that are not in hand as specimens and that are not exactly like the specimens in hand. The specimens are, then, a sample from which group or population characters are estimated and this is the only proper sense in which they are the basis of a species in taxonomy. The members of this sample for estimation are thus "types" in one of the three senses in which that word is currently used.

The second basic operation of taxonomy is comparison, either comparison of species or identification of specimens, which is a different aspect of the same operation. Having formed an estimate from a sample, or a species concept from certain specimens, the purpose now is to see whether specimens hitherto excluded from the sample fall within the boundaries of that estimate and concept. If it is decided that they do or should (that the two species are synonymous or that the specimens being identified belong to a given species), then these additional specimens should be placed in the sample and one should return again to the first operation and redefine the species, because any change in the sample permits and demands a change in the inferences based on it. The change may be either a shift of boundaries or a narrowing of the limits of estimation, bringing them closer to the scope of the real species, or it may be and usually is both.

The sample taken at the beginning of this operation as representative of the species is a standard of comparison and its members are thus "types" in the second of the three senses. The procedure here outlined is, however, importantly unlike the usual classical "comparison with types." That procedure usually involved taking a specimen or a series of specimens, "types," as inflexibly standard and typical for a given species and making subjective judgment as to the degree of difference from them to be permitted to other specimens referred to the same species. It is a distinctly different procedure to take all the available material hitherto referred to the species and to determine whether the addition of other specimens would still give a sample such that an orderly group concept developed from it by controlled inference would correspond with the idea of a species. It is this procedure that is now advocated as the proper method of comparison and identification in modern taxonomy.

According to this principle of taxonomy, any identification or reference of specimens to a species makes possible and implicitly or explicitly involves redefinition of that species. Two workers seldom base their concepts of a species on exactly the same suites of specimens. Their concepts therefore will seldom be exactly the same and since these concepts, and not the natural or real groups, are the species of taxonomy, it follows that two original workers seldom use the same species. Taxonomy might thus disintegrate into an almost mystic chaos of individual systems were it not for a working agreement to which all taxonomists subscribe. This is that however much their bases for inference may differ in all other respects they will always include one or more specified specimens. Individual concepts can be conveyed in more or less concrete terms and this communication permits different students to bring their concepts into some agreement, or at least nondisagreement, as far as consistent with their general frames of reference. By a convention essential to this communication of ideas, it is agreed that the application of a certain name to a species is an implicit guarantee that specimens attached in a legal way to that name are included in the sample on which the species concept of the author now using the name is based.

A third basic operation of taxonomy is thus determining whether or not certain published specimens with names attached to them fall into a given specific concept, selecting the correct name if two or more such name-bearers fall into a single group concept, and coining a new name and attaching it to a particular specimen if no previous name-bearer is included in the concept. These specimens with names attached to them are also called "types," and it should be apparent that "types" in this sense are being used in a very different way from the two sorts previously mentioned.

Most taxonomy has hitherto assumed that the same specimens, the "types," adequately serve all three of these functions. Some taxonomists apparently feel not only that the single set of specified "types" can do this triple duty but also that it must do so. The view of taxonomy as a system of group concepts based on inferences about populations from samples, a view that is rapidly gaining ground and to which I strongly adhere, is decisively incompatible with this use of "types." According to this principle the specimens used as "types" in the three different ways not only need not be, but also cannot be, the same if proper scientific methods are followed.

In the original proposal of a species, the inference involved should be based on all the available specimens that are then considered as belonging to the species and on all of them equally. No extra weight should be given to any one specimen or small group of specimens that are shown or believed to be more "typical," nearer a norm or point of central tendency for the species. Every specimen, however "aberrant" or deviant from the norm, necessarily has equal weight in the correct drawing of inferences from the sample, providing that the specimen can properly be placed in the sample. In other words, every specimen that is referred to a species by its original describer should be used by that describer as part of his basis for the species and none should be singled out as "type" and considered as contributing more to the concept than does any other specimen.

In order to have a necessary word to use in this discussion, the matter of terminology may here be anticipated on one point. It would be possible to extend the term "types" to include all the specimens used by the original author, not "holotype" and "paratypes" etc., but simply "types," because all are coordinate in this use. This would, however, lead to endless confusion, as will become still more apparent in the following paragraphs. I dislike adding to terminology, but I do not see how to convey this idea clearly and succinctly without the use of some single word not liable to serious confusion with the quite different concepts hitherto included under the names "types," and I therefore propose the term "hypodigm" (pronounced hy'-podim, from the Greek hypodeigma, "token, example"). All the specimens used by the author of a species as his basis for inference, and this should mean all the specimens that he referred to the species, constitute his hypodigm of that species.

In a subsequent comparison or identification, the basis of comparison is not correctly a "type," in any restricted sense, but a hypodigm. Even if only one specimen was known to the original author, it should be used by him and for subsequent comparison as a hypodigm and not as a type in any other sense. Here "type" and "hypodigm" necessarily coincide in referring to the same specimen, but they do not become synonymous unless "type" be redefined. The type is an isolated object. The hypodigm, whether it include one specimen or a thousand, is a sample from which the characters of a population are to be inferred. In one case the mental process of an identifier is, in essence, "This specimen does not appear to differ notably from that (the type), so I will call it by the name of that." In the other, and scientifically correct, procedure, the identifier might say, "From this hypodigm, even though it includes only one specimen, I infer certain broad limits within which the boundaries of a real group probably lie. Since the specimen to be identified is within these broad limits, it may belong within the boundaries of the real group. As a working hypothesis I will place the two specimens together and use them as an enlarged hypodigm on which I can base certain narrower limits around the unknown real boundaries of the natural group that I am calling a species."

Each subsequent identification adds to the material on which the concept of the species can and should then be based and with which the next comparison should be made. These subsequent materials thus become "types" of the species as it is constantly redefined and as bases for comparisons, but to call them so would obviously violate the now accepted meaning of the word "type." They become parts of the hypodigm, which is properly subject to frequent change. Every specimen placed in a hypodigm at a given time by any one author, whether at the time of original publication or centuries later, is coordinate with every other specimen then in the hypodigm. To distinguish principal, original, or primary from subsidiary, subsequent, or secondary hypodigms, as has been done for "types," would be to miss the significance of the whole procedure as it is correctly carried out on a sampling basis. A unique original specimen on which a species was first based has no more authority or weight for forming a valid concept of that species than has any one of a hundred specimens subsequently referred to that species and is no better as a standard of comparison.

Since data not imparted are not part of science as an impersonal body of fact and theory, specimens cannot be recognized as parts of a hypodigm unless publication has been based on them. With this exception every specimen definitely referred to a species is or has been part of its hypodigm. It is advisable that every worker specify the members of his hypodigm, some of which may be known to him at first-hand and others through the literature based on them. Commonly this can be done by stating that the current hypodigm is that of the original describer, or of some other reviser, plus and minus certain specimens. It is, of course, frequently necessary for a reviser to drop from a hypodigm specimens formerly placed in it. The hypodigm, the real basis for the species, is thus never properly a specified and static list of specimens placed in the species at any particular time or labelled as any particular sort of "type," but is and should be a fluid and constantly changing thing. In historic review it is proper to refer to what the hypodigm was for a particular author at a particular time, and this is generally far more significant than to state what that author accepted as "type."

From the point of view of strict comparison, a hypodigm is an indirect means of comparing with the real species that it represents. A species cannot be directly observed or compared; the real species because it is never available as such, the taxonomic species because it is merely a mental concept based on observations, not a thing subject to observation. When a single specimen or limited group of specimens are singled out as "types" to be standards of comparison, some observation or assumption must be made as to their position in a hypodigm. The classic "comparison with the type" is properly considered only as a means of indirect comparison with a hypodigm, itself a means of indirect comparison with the real species. Surely every effort should be made to modify a method so full of indirection. Even if the degree of indirection cannot be much reduced, there is no mystic virtue in "types," as such, that makes them any better for comparison than would be any other member of a hypodigm. If one or a few specimens must be singled out, they should be as nearly as possible near an average point in the hypodigm, and in practice "types" very frequently are worse in this respect than are many simple "referred specimens."

For the third "type" function, that of name-bearer and anchor, a specimen that gives a species concept a name and that is by the name used guaranteed to be included in the hypodigm of the species, the requirements are very different. This convention requires a type in the classic sense of some specimen designated as such at the outset, once and for always.

No one can ever guarantee that two specimens will always be placed in one hypodigm, so that it is highly desirable that only one name-bearer be designated. If two or more cobearers were originally given to a name, a subsequent reviser has the duty of cutting all but one of them adrift and leaving each name fastened to only one object.

Within the hypodigm where it is currently placed, a name-bearer may be a very poor basis for inference and a worse means of comparison, but this does not matter. It does not matter, that is, as long as no one insists that the name-bearer must be used exclusively or must be given added weight and authority for these other two purposes. To the extent that the confused idea of "types" so widespread in the past has considered these to be certain specimens of particular importance for all three purposes, it should now be apparent that these ideas and the terminology and practices based on them require revision.

Source

(1940) Types in Modern Taxonomy, *American Journal of Science*, Vol. 238, pp. 413–431.

Another reference

STOLL, N.R., R. PH. DOLLFUS, J. FOREST, N.D. RILEY, C.W. SABROSKY, C.W. WRIGHT, and R.V. MELVILLE, editors (1964) *International Code of Zoological Nomenclature*, International Trust for Zoological Nomenclature, London.

Taxonomic Linguistics

Comment

For some years the Wenner–Gren Foundation of New York also maintained a conference center in Burg Wartenstein, an ancient castle, comfortably modernized, at the border of the Austrian Alps. The summer meetings there, which alas! have had to be discontinued, were not only delightful for the participants but also highly productive in the exchange and increase of knowledge and insight in many aspects of anthropology. A meeting in 1962, arranged by Professor Washburn of the University of California and managed by the late Dr. Fejos of the Wenner–Gren Foundation, was devoted to "Classification and Human Evolution." I was asked to attend and to speak in a general way on the meaning of taxonomic statements and more specifically on the interpretation of primate phylogeny in taxonomic terms.

The text following here is a version of the first part of my contribution to that conference. The participants prepared preliminary drafts which were circulated in advance, and the present text is slightly modified from that draft. It was later rewritten to include mention of results of the conference as a whole, and that different version was printed in the volume cited at the end of this section of this book. The bulk of the final version went into considerable detail about the phylogeny and classification of man and some other primates and is not here reprinted.

A linguistic approach to the terminology of taxonomy and the nomenclature of classification was a novelty. It was well received at the conference and has influenced subsequent work by anthropologists. As it appeared in a strictly anthropological context, it has received little attention from nonanthropological professional taxonomists. I have not had occasion to follow this version with one for the broader audience, and I hope that this somewhat modified reprint will lead to the further attention, whether favorable or not, that I think this approach merits.

Text

Taxonomic language involves not only a very large number of different designative words (names, terms) but also several different *kinds* of designations. The things or concept designated by these words, technically their referents, are also of different kinds, and the meanings or semantic implications are likewise diverse. It is therefore essential that they be clearly distinguished. One way to do this is to consider the main operations involved in classification and the points or levels where special designations are required, as shown schematically in Fig. 4–22.

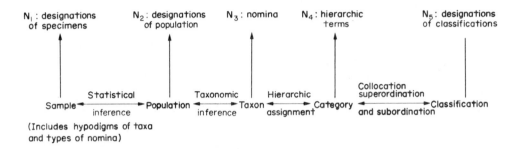

FIG. 4–22. Schema of processes (arrows), name sets (N), and referents (capitals) in linguistic analysis of the process of classification and nomenclature. Further explained in the text.

The process starts with observation of the specimens in hand, the objective materials. The specimens studied and believed to be related in some biologically relevant way are a *sample*. If they are believed to represent a definite taxon (as determined at another level of inference), they constitute a *hypodigm*. Unequivocal designations of the specimens must refer to them as concrete, discrete objects; they are not designated by any name of the population or taxon to which they are supposed to belong. The ideal designation, practically universal in zoology but unfortunately not in anthropology, is by a collection or repository symbol and a catalogue number uniquely associated with each specimen. This is one kind of designation, one set of names (symbols of some sort, not necessarily or usually in words), and may be called the N_1 naming set.

Observations and specimens, no matter how numerous, have no scientific significance purely per se. They acquire significance only when they are considered as representative of a larger group, or population, of possible observations or of individuals united by some common principle or relationship. The population may be abstract, for instance as symbolized in the equation for gravitation, applicable to a potentially infinite number of events but derived from a finite series of experimental observations. In zoological taxonomy the population is finite and concrete: a set of organisms existing (now or formerly) in nature. The existence and characteristics of that population are inferred from the sample drawn (we hope at random) from the population. The methods of inference are statistical by definition, which does not mean that any particular procedure of mathematical statistics is necessarily used although, of course, that is often appropriate and useful. A population is obviously not the same as the specimens actually studied, a sample drawn from the population.

At the next step in the process, all populations belong to taxa and all taxa are composed of populations. However, the two are not necessarily coextensive. It is often necessary to recognize and designate a local population that is a part of a taxon but does not in itself comprise a whole taxon. For some populations a different set of names or symbols, N_2 may therefore be required. Populations are in fact sometimes given distinct designations in zoological systematics, commonly by specification of their geographic location, but there is no established and uniform system. It may be sufficient to designate a population either as that from which a given sample was drawn (hence by extension of an N_1 designation) or as identical with that of a given taxon (hence by an N_3 designation).

A taxon is a group of real organisms recognized as a formal unit at any level of a hierarchic classification (Simpson, 1961, which see also for definitions and more extended discussions of hypodigms, categories, and hierarchies). A taxon is therefore a population, although the overall population of one taxon may include many distinct populations of lesser scope. A taxon is created by inference that a population (itself statistically inferred from a sample which now becomes a hypodigm) meets a definition adopted for units in an author's classification. The set of designations for taxa, N_3 names, are those of formal, technical zoological nomenclature, e.g. *Homo*, the name of a taxon in primate classification. The word "name" is used in many different ways, both in the vernacular and in technical discussion, and this has engendered confusion. I propose that technical Neo-Linnaean names in the N_3 set to be called *nomina* (singular, *nomen*). Vernacular names ("lion," "monkey," "Neanderthal man") are in the N_3 set if they designate taxa, but they are not nomina.

Each taxon is assigned to (considered as a member of) a category, which has a defined rank in hierarchic classification. A category is a set, the members of which are all the taxa placed at a given level in such a classification. Categories are distinct from taxa, do not have populations as members, and are not represented by samples. They have their own set of names, N_4, which are the relatively few terms applied to levels of the Neo-Linnaean system: basically phylum, class, order, family, genus, and species, with various combinations in super-, sub-, and infra-, and occasionally such additional terms as cohort or tribe.

Finally the various taxa of assigned categorical rank are collocated, superordinated, and subordinated among themselves and so form a hierarchic classification. This is done in terms of the N_3 (nomina) and N_4 (hierarchic terms) names. The added implications are conveyed less by nomenclatural than by topological means, primarily by arrangement and not consistently by verbal or related symbolization. Designations of classifications, N_5, are normally bibliographic references to their authors and places of publication.

What, now, are the meanings or implications of the various sets of designations? N_1 designations refer to particular objects. They imply only that a given specimen exists. They assure that when the same designation is used, the same object is meant. N_2 and N_3 designations both refer to groups that are considered to be populations related in some way. An author using such designations must make clear, explicitly or implicitly, the kind of relationship he has in mind. In modern zoology unless some other usage is definitely stated, it is generally understood that the relationship is genetic, that is, that it reflects evolutionary relationships. Concepts of what constitutes evolutionary relationships, how they are to be determined, and how reflected in classification become difficult and complex, but that is a different point.

Besides the implication that a population, usually genetic in relationship, is designated, nomina, N_3 names, further imply that the unit designated is given a definite rank in classification, that it is associated with an N_4 term. Under the International Code the forms of some nomina reflect the categorical rank of the corresponding taxa. For example, nomina ending in -idae (e.g. Hominidae) name families, and italicized, capitalized single words (e.g. *Homo*) name genera.

Most nomina, however, lack implications as to superordination, and *none have any implications beyond those mentioned*. For instance, nomina have no implications as to relationships among taxa at the same categorical level (e.g. *Homo* and *Tarsius*) or among taxa at any levels with etymologically distinct names (e.g. *Gorilla* and Pongidae). Further implications, which may be numerous and intricate as will be illustrated later, are inherent in the arrangement of nomina in a classification and not in the nomina themselves.

It must be emphasized that one of the greatest linguistic needs in this field is for clear, uniform, and distinct sets of N_1 and N_2 designations, applied to specimens and to local populations as distinct from taxa. Just what form such designations should take is a matter for proposal and agreement among those directly concerned with the specimens and their interpretation. It suffices here to stress that they *must not* have the form of Neo-Linnaean nomina.

Men and all recent and fossil organisms pertinent to their affinities are animals, and the appropriate language for discussing their classification and relationships is that of animal taxonomy. When anthropologists have special purposes for which zoological taxonomic language is not appropriate, they should devise a separate language that does not duplicate any of the functions of this one and that does not permit confusion with its forms. There is, I believe, no reason for use of an additional language when what is being discussed is in fact the taxonomy of organisms. This language has been developed over a period of hundreds of years by cumulative experience and thought and has been thoroughly tested in nonanthropological use. It is admittedly imperfect, but for its purpose it is the best instrument available. Its imperfections call rather for improvement than for replacement.

It is notorious that hominid nomenclature, particularly, has become chaotic. It is ironical that some of those who have most complained of the chaos have been leading contributors to it. A recent proposal that an international commission be formed to deal with the chaos refuses to recognize the appropriate code and the appropriate commission already set up. The author then proceeds to compound the confusion that he condemns.

Insofar as the chaos is merely formal or grammatical, it could be cleared up by knowledge of and adherence to the International Code of Zoological Nomenclature, supplemented, if necessary, by whatever action might be proposed to and endorsed by the International Commission for Zoological Nomenclature. Much of the complexity and lack of agreement in nomenclature in this field does not, however, stem from ignorance or flouting of formal procedures but from differences of opinion that cannot be settled by rule or fiat. Insofar as the chaos is due to faulty linguistics rather than to zoological disagreements, it stems either from ignorance or from refusal to follow rules and usages. This must be almost the only field of science in which those who do not know and follow the established norms have so frequently had the temerity and opportunity to publish research that is, in this respect, incompetent.

An overt reason sometimes given for refusal to follow known nomenclatural norms is that some nomen is, in the opinion of a particular author, inappropriate. For example, some choose to rename *Australopithecus* as *Australanthropus*, thus adding another objective synonym to the chaos, on the grounds that the Greek *anthropos* more nearly expresses their opinion as to the affinities of the genus than does *pithekos*. The argument is completely irrelevant. *Australopithecus* does not mean "southern ape." Its meaning (defined by its referent) is simply the taxon to which the nomen was first attached and to which it was the first nomen attached. *Palaeolumbricus* or *Jitu* would have served just as well. The generic nomen does not, in itself, express any opinion as to the affinities of the taxon, and if nomina were changed in accord with every shade of opinion on affinities the chaos would be even worse than it is.

Another reason for the chaos is the previously mentioned failure to develop and use consistently different designations for specimens, populations, and taxa, that is, distinct N_1, N_2, and N_3 name sets. A truly eminent anthropologist insisted on using the (N_3) nomen *Sinanthropus pekinensis* for specimens and a population although he concluded that his nomen does not designate a *taxon* specifically distinct from *Pithecanthropus erectus* or indeed from *Homo sapiens*. The example is far from unique.

Probably no one has ever admitted this, but it seems almost obvious that nomina (N_3) have sometimes been given to single specimens just to emphasize the importance of a discovery that could and should have been designated merely by a catalogue number (N_1). Of course no two specimens are alike, and it is always possible to fulfill the formal requirement that ostensible definition of a taxon must accompany proposal of a nomen. However, and again I would say obviously, the "definition" has often been only a description of an individual "type" with no regard for or even apparent consciousness of the fact that taxa are *populations*. This is not just a matter of exaggerating the taxonomic difference between specimens. It is a much more fundamental misunderstanding of what taxonomy is all about, of what nomina actually name. It is a relapse into pre-evolutionary typology, from which (I must confess) even the nonanthropological zoologists have not yet entirely freed themselves. Nomina have types, but not in the old typological sense. The types are not the referents of the N_3 nomina but are among the referents of N_1 designations. The referents of nomina are taxa—certain kinds of populations.

It is of course also true that the significance of differences between any two specimens has almost invariably come to be enormously exaggerated by one authority or another in this field. Here the fault is not so much lack of taxonomic grammar as lack of taxonomic common sense or experience. Many fossil hominids have been described and named by workers with no other experience in taxonomy. They have inevitably lacked the sense of balance and the interpretive skill of zoologists who have worked extensively on larger groups of animals. It must, however, be sadly noted that even broadly equipped zoologists often seem to lose their judgment if they work on hominids. Here factors of prestige, of personal involvement, of emotional investment rarely fail to affect the fully human scientist, although they hardly trouble the workers on, say, angleworms or dung beetles.

It is not really my intention to read an admonitory sermon to the anthropologists. You are all well aware of these shortcomings—in the work of others. I must pass on to matters more positive in value.

The undue proliferation of specific and generic nomina is in part a semantic problem. The proposal of such nomina is rarely accompanied by an appropriate definition of the categories (as distinct from the taxa) involved, but ascribing specific or generic status to slightly variant specimens can be rationalized only on a typological basis. Whether consciously or not, taxa are evidently being defined as morphological types and statistical-taxonomic inferences from hypodigm to population to taxon (see Fig. 4–22) are being omitted. But in modern biology taxa are populations and the following two nonconflicting definitions of the species are widely accepted:

Species are groups of actually or potentially interbreeding populations, which are reproductively isolated from other such groups.

An evolutionary species is a lineage (an ancestral–descendant sequence of populations) evolving separately from others and with its own unitary evolutionary role and tendencies. (Quoted from Simpson, 1961, where sources are cited and the definitions are further discussed.)

The naming of a species either should imply that the taxon is believed to correspond with one or both of those definitions or should be accompanied by the author's own equally clear alternative definition.

Evidence that the definition is met is largely morphological in most cases, especially for fossils. The most widely available and acceptable evidence is demonstration of a sufficient level of statistical confidence that a discontinuity exists *not* between specimens in hand but *between the populations inferred from those specimens*. The import of such evidence and the semantic implication of the word "species" are that populations placed in separate species are either

(1) in separate lineages (contemporaneous or not) between which significant interbreeding does not occur, or

(2) at successive stages in one lineage but with intervening evolutionary change of such magnitude that populations differ about as much as do contemporaneous species.

In dealing with the incomplete fossil record the information at hand commonly cannot establish the original presence or absence of a discontinuity. Allowance must be made for probabilities that further discovery will confirm or confute the existence of an ostensible discontinuity. Those probabilities depend on various circumstances. If populations are approximately contemporaneous, only moderately distinctive, and separated by a large geographic area from which no comparable specimens are known, there is considerable possibility that discovery of intervening populations would eliminate discontinuity. That is, for example, the situation regarding the original hypodigms of *Pithecanthropus erectus* and *Atlanthropus mauritanicus*. In my opinion the possibility that the Trinil population and the Ternifine population belong to the same species is such that different specific (*a fortiori*, generic) nomina are not justified at present.

If, on the other hand, populations being compared are of markedly different ages, decision to give them different specific nomina should depend on judgment whether such nomina would be justified if it turned out that they belong in successive segments of the same lineage. That would apply, for example, to the Mauer population as compared with the late Pleistocene European neanderthaloid population, and I should think would justify different specific nomina in this example. Still a third situation arises when samples indicate populations that were approximately contemporaneous and living in the same region (synchronous and sympatric) as may be true, at least in part, for the Kromdrai, Swartkrans, Makapan, and Sterkfontein populations. In such cases allowance hardly has to be made for possible discoveries of populations living at other times and in different places. The degree of statistical confidence generated by the samples actually in hand may be taken as definitive of the probability of an original discontinuity, for instance between *Australopithecus africanus* and *A. robustus*.

The category genus is necessarily more arbitrary and less precise in definition than the species. A genus is a group of species believed to be more closely related among themselves than to any species placed in other genera. Pertinent morphological evidence is provided when a species differs less from another in the same genus than from any in another genus. When in fact only one species of a genus is known, that criterion is not available, and judgment may be based on differences comparable to those between accepted genera in the same general zoological group. There is no absolute criterion for the degree of difference to be called generic, and it is particularly here that experience and common sense are required.

It must be kept in mind that a genus is a *different* category from a species and that it is in principle a *group* of species. Much of the chaos in anthropological nomenclature has arisen from giving a different generic nomen to every supposed species, even some clearly not meriting specific rank. In effect no semantic distinction has been made between genus and species, and indeed the number of proposed generic nomina for hominids is much greater than the number of validly definable species. Monotypic genera are justified when, and only when, a single, isolated known species is so distinctive that the probability is that it belongs to a generic group of otherwise unknown ancestral, collateral, or descendent species. No one can reasonably doubt that this is true, for example, of *Oreopithecus bambolii* and that in this case the (at present) monotypic genus is justified. It is, however, hard to see how the application of more than one generic name to the various presently known australopithecine populations can possibly be justified, whatever the specific status of those populations may be.

As most biologists understand modern taxonomic language, its implications are primarily evolutionary, but there is some persisting confusion even among professional taxonomists. It is not possible for classification directly to *express*, in all detail, opinions either as to phylogenetic relationships or as to degrees of resemblance. As a rule with important exceptions, degrees of resemblance tend to be correlated with degrees of evolutionary affinity. Resemblance provides important, but *not the only*, evidence of affinity. Classification can be made consistent with, even though not directly or fully expressive of, evolutionary affinity, and its language then has appropriate and understandable genetic implications. Classification cannot, at least in some cases, be made fully consistent with resemblance, and any implications as to resemblance are secondary and not necessarily reliable. These relationships can be explored by consideration of several hypothetical models or examples, set up so as to be simplified parallels of real problems in the use of taxonomic language to discuss human origins and relationships.

Classification and taxonomic discussion of related but distinct contemporaneous groups, such as the living apes and living men, involves a pattern of evolutionary divergence. That will first be discussed by means of a model. Discovery of related fossils almost always complicates the picture by revealing other groups divergent from both of those primarily concerned. It may, however, also reveal forms that are ancestral or that are close enough to the ancestry to strengthen inferences about the common ancestor and the course of evolution in the diverging lineages. In general the characters of two contemporaneous groups as compared with their common ancestry will tend to fall into the following classes, exemplified by characters of recent Pongidae and Hominidae:

(A) Ancestral characters retained in both descendent groups: e.g. absence of external tail, pentadactylism, dental formula.

(B) Ancestral characters retained in the first descendent group but divergently evolved in the second: e.g. quadrupedalism, grasping pes.

(C) Ancestral characters retained in the second but divergent in the first group: e.g. undifferentiated lower premolars.

(D) Characters divergently specialized in both: e.g. brachiation versus bipedalism.

(E) Characters progressive but parallel in both: e.g. increase in average body size.

(F) Convergent characters. I know of none between pongids and hominids, a fact which (if it is a fact) greatly simplifies judgment as to their affinities.

Different numbers of characters will fall into different categories. For instance in pongid–hominid comparison there are certainly many more A characters than any others and more B than C characters. (The given example of a C character is dubious.) Many characters do not simply and absolutely fall into one category or other. Retention of ancestral characters is usually relative and not absolute; some changes generally occur and "retained" usually means only "less changed." In constructing the simplest possible model on this basis, further simplifying postulates are that characters evolve at constant rates and that characters in the same group (e.g. D or E) evolve at the same rates. Those postulates are certainly never true in real phylogenies, and more realistic but also much more complicated models can be constructed by taking varying rates of evolution into account. The simplest possible limiting case, although unrealistic in detail, nevertheless more clearly illustrates valid and pertinent matters of principle. Such a model, analogous to pongid–hominid divergence, is illustrated in Fig. 4–23. Numbers preceding the category designations symbolize relative numbers of characters in the corresponding categories. Exponents symbolize progressive change: a–b–c, or in a different direction x–y–z. It is assumed that in this example there are no F (convergent) characters. Roman numerals represent taxa: IV and V the two contemporaneous groups being compared, and I their common ancestry, ancestral to IV through II and to V through III.

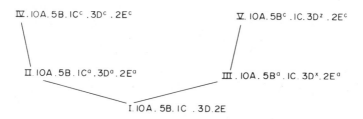

Fig. 4–23. A conceptual model of simple evolutionary divergence. The symbols are explained in the text.

From such data a comparison matrix can be formed. There are more sophisticated ways of doing this but for present purposes a simple and sufficient method is to tabulate step differences between taxa. Change from C to C^a, for instance, is one step and from C^a to C^c is two more. These are multiplied by the number of characters in the category, 1 for C characters. The matrix for the model in Fig. 4–23 is given in Table 4–1. In this form of comparison, the smaller the number the greater the similarity. In this model I and II are most and IV and V least alike.

Let us suppose now that classification were to be based *entirely* on degrees of resem-

TABLE 4–1

Comparison Matrix for Data of the Model in Fig. 4–20

	I	II	III	IV	V
I	0	6	9	18	29
II	6	0	12	12	32
III	9	12	0	24	19
IV	18	12	24	0	36
V	29	32	19	36	0

blance, as has been proposed by some taxonomists, and that classificatory language was therefore understood to be directly and solely expressive of resemblance. In building up higher taxa one would of course start by uniting I and II. If I and II are species, they would be placed in one genus; if genera, in one family. The maximum difference within the higher taxon would be 6. If no greater difference were allowed, all other lower taxa, III, IV, and V, would have to be placed in separate, monotypic higher taxa, an arrangement with nearly minimal significance, indicating no more than the close resemblance of I and II. If a difference of 12 were allowed in the higher taxon, III would be united with I and II, but IV should now also go with II, from which its difference is also 12. However, a taxon including IV and I would have to allow a difference of 18 and one including IV and III a difference of 24. But now V must also be added, for its difference from III is only 19. Thus *all* the lower taxa must go in a single taxon of next higher rank, an arrangement that indicates nothing of resemblances or relationships among any of those taxa. Insertion, or in actual examples discovery, of additional taxa, say between II and IV, would only compound the difficulties and lead still more inevitably to equally unsatisfactory alternatives.

I believe that the conclusion from the model is quite general for analogous real cases. In such situations the use of classificatory language as direct expression of degrees of resemblance commonly tends to produce one of two extremely inexpressive results: (1) one higher taxon includes the two most similar lower taxa and all other higher taxa are monotypic; or (2) one higher taxon includes all the lower taxa, no matter how numerous.

Now let us agree that classificatory language is to have primarily evolutionary significance. For the moment, degrees of resemblance need not be considered at all. It is clear from consideration of characters in categories B, C, and D that II can be ancestral to IV but not to III or V, and that III can be ancestral to V but not to II or IV. In actual instances the conclusions are neither so simple nor so obvious, but probabilities are readily established by the same categories of evidence. On this basis, I and IV can be placed in one higher taxon and III and V in another of the same categorical rank. That arrangement expresses the opinion, postulated as true in the model, that II and IV are phylogenetically related to each other and that III and V are also related in more or less the same way and degree. The arrangement is also consistent with but does not express the opinion, also postulated as true, that II is ancestral to IV and III to V.

In completion of this arrangement there are two alternatives as regards I. It could be placed in a third higher taxon ancestral to both of the two already formed, or it could be placed in the same higher taxon as II and IV, because it is phylogenetically closer to II than to III. Degree of resemblance here enters in as evidence for the latter inference.

Those are not the only classifications that would be consistent with the postulated evolutionary history. It would also be consistent to put I, II, and III in one higher taxon and IV and V in two others, or I, II, III, and IV in one and V in another. The implications on affinity would be somewhat different in each case but not conflicting: all are consistent with the postulates of the model. Choice would depend in part on what implications one wanted especially to bring out, since not all can be expressed in one classification. It would also depend on other considerations such as not changing previous classifications unnecessarily and conveying as much significant information as possible. (The last alternative mentioned above is the least informative.)

The model also illustrates the tendency, which is open to exception, for degree of resemblance to correlate with nearness of common ancestry. II and III are nearer their common ancestry than are IV and V, and they resemble each other more closely. The same is true of III and IV or of II and V as against IV and V. Such relationships are not directly implicit in the classification, but they are important in arriving at the judgments of affinity that are implicit in it.

Another important point illustrated in the model is that II and III resemble each other much more closely than III resembles its descendent V. It is realistic to expect an early—say Miocene—ancester of *Homo* to be more like an ancestral ape than like modern man. It is unrealistic to expect the Miocene ancestors of either (or both) groups *necessarily* to have any of the specialized features that are diagnostic between *recent* members of the two families.

Source

As noted above, this text is not a direct reprint from the different previously printed version but is derived and modified from the same source. That printed version is:

SIMPSON, G.G. (1963) The meaning of Taxonomic Statements, in *Classification and Human Evolution*, (S.L. WASHBURN, editor) Aldine, Chicago, pp. 1–31. (The present text approximately parallels pp. 1–12.)

Other references

STOLL *et al.* (1964), cited on p. 164, and SIMPSON (1961), cited on p. 159, are again relevant here.

5

Some Bits Of Biometry

BY THE 1930s quite sophisticated biometrical and general statistical methods were being widely used in such diverse fields as demography, agriculture, genetics, growth, anthropology, and psychology. In that respect zoology, both neo- and paleo-, was lagging behind. The usual zoologist of that period not only neglected but also rejected any but the most elementary mathematics. Many, or indeed most, of the systematists were still typological thinkers by whom the description of a single "type" specimen and with a few of its linear measurements were taken as representative of a species. Even those who were beginning to think nontypologically and in terms of variable populations often failed to recognize that an essential apparatus for such studies must include biometrical statistics both in concepts and in measures.

The change in attitude not only advisable but also necessary for further progress in zoology was recognized and, I believe, accelerated by a book on quantitative zoology by my wife Anne Roe and me, published in 1939, later (1960) revised and rewritten with the further co-authorship of Richard Lewontin.

My first extensive use of statistical methods was in the 1937 memoir cited on p. 96 of the present work. There had been a few previous paleontological applications of statistics, for example by F.H. Frost in 1927 in an appendix to a paper on a fossil bird and by Helga Pearson, daughter of the eminent biometrician Karl Pearson, in 1928 in a monograph on Chinese fossil pigs. Still these methods had not been generally adopted and in fact were still viewed with some distaste by most paleontologists. It was thus still unusual to state that "throughout this work, wherever they proved useful, statistical methods have been employed." Reference was made to R.A. Fisher's somewhat esoteric work and also to "a paper soon to be published (Simpson and Roe)." The "paper" was not published as such, and in our hands it soon grew into the book mentioned above. Because the essentials were fairly well covered in that book and its successor, I have had little occasion to write separate papers strictly on biometric methods.

The populational and statistical point of view is evident in several of the items included in other sections of this book. Here I include only quotation from one explicit biometric example, among many possible, and two innovative and specifically biometric proposals. In fact neither one of these proposals has been widely followed, and that is one reason why I do include them here. I believe that they may still be worth some thought.

I hardly need to add that almost all well-trained zoologists are now well trained in many biometrical procedures.

References

SIMPSON, G.G., and A. ROE (1939) *Quantitative Zoology: Numerical Concepts and Methods in the Study of Recent and Fossil Animals*, McGraw–Hill, New York and London.
SIMPSON, G.G., A. ROE, and R.C. LEWONTIN (1960) *Quantitative Zoology*, Revised Edition, Harcourt, Brace, New York and Burlingame.

An Application of Statistical Thinking and Methods

Comment

Early in the present century the American Museum of Natural History brought together a large collection of extinct Cuban ground sloths, mostly of a single genus, *Megalocnus*. W.D. Matthew, then Curator of Fossil Mammals at that Museum, made a preliminary study of them and started a more extended manuscript but had not finished it before he left the Museum in 1927. A brief publication extracted from his manuscript was published in 1931, the year after Matthew's death, but the complete study of the collection was in abeyance until 1951 when the Brazilian paleomammalogist Carlos de Paula Couto was working at the American Museum and added this to his other researches.

Paula Couto asked me to make a biometric analysis of the supposed species of *Megalocnus*. That analysis was published as an appendix to the monograph and is here reprinted in abbreviated form to exemplify the application of statistical concepts and methods to a particular problem. The accompanying figure also illustrates a pictorial representation of size classes and ratios of two dimensions considered simultaneously.

Bradypus and *Choloepus*, referred to in the following text, are the two genera of living tree sloths.

Text

Matthew worked before the establishment of the new systematics, with its emphasis on populations and distribution of variation rather than on types. He was, nevertheless, one of the forerunners of this movement, and even his early work shows an awareness of wide variation as a usual characteristic of the natural populations that are species. He was, moreover, one of the first to insist that two or more closely similar or related but distinct populations are rarely found in the same place at the same time. A paleontological corollary, which was explicitly drawn by Matthew and perhaps for the first time, is that in a collection of unified origin (from a single fossil quarry, for instance) congeneric animals will generally be of a single species, or, if they are of more than one species, the discontinuity between the groups will be large and evident. In such a collection, then, in the absence of fairly obvious discontinuity, the variation within a genus should usually be taken as intraspecific. The variation of the species is given by the data, which are not to be subjected to *a priori* ideas as to how much variation is allowable. There are certainly some exceptions to this rule, but there is now little doubt that it is the rule.

That principle is here restated and the fact that we owe it to Matthew as much as to any one person is emphasized because of the surprising fact that Matthew himself did not apply it to the Cuban sloths. In the single genus *Megalocnus* he defined three species, and in one of them two subspecies, as all present together in the apparently quite unified *casimba* fauna. Paula Couto has (in my opinion, quite properly) reduced one of those supposed species (*M. "junius"*) to synonymy. He has tentatively recognized the other three taxa of Matthew, but has drawn attention to the possibility that they also may all be synonymous. That Matthew himself had doubts is evident in his letter to de la Torre, quoted on a preceding page. He there noted that his four *provisional* (my italics) taxa in *Megalocnus* "may be partly due to age and individual differences, although they can hardly be all referred to one species, for the range in size is over 300 per cent, far more than in a series of different aged skulls of *Bradypus and Choloepus*." Matthew's "300 per cent" is here an *a priori* criterion and hence contrary to his own principle, although no one would quarrel with it if growth and sexual dimorphism were definitely excluded as significant factors. The comparison with the recent *Bradypus* and *Choloepus* does not sufficiently account for the possible influence of growth, because (a) we do not know (and it is indeed improbable) that the examined recent specimens covered as wide an age range as the specimens of *Megalocnus*, and (b) we do not know (and again it is improbable) that growth changes in available dimensions are comparable in *Bradypus* or *Choloepus* and in the much larger and quite differently proportioned *Megalocnus*.

It must thus be admitted that there is some weakness in Matthew's conclusions on this point. Matthew was aware of possible weakness here, and if he had been able to revise and complete his manuscript he might well have changed it, at least to the extent of reducing the supposed taxa in *Megalocnus* from four to two. Paula Couto, equally conscious of the problem, has asked me to make an independent check by biometric methods, although of course also aware that such methods cannot by themselves establish a definitive solution.

In the collections here studied specimens that can be shown beyond doubt to be completely adult (crania, vertebrae, and long bones with sutures completely closed or epiphyses fully fused) are much too few to permit reliable statistical discrimination of possibly separate populations. It is, nevertheless, noteworthy that observed differences in dimensions of such specimens do not even remotely approach Matthew's figure of 300 per cent (which therefore was not based on dimensions known to be of full-grown individuals). No adult specimens have been referred to the smallest two supposed species (*M. ursulus* and *junius*)—a fact highly suggestive in itself. The best comparisons available within *M. rodens* are for length of humerus (eight specimens) and length of femur (nine specimens). For the humerus the ratio of largest to smallest is 1.39 and for the femur 1.53. These ratios are large for single populations of mammals, but not impossibly so, especially if there is any sexual dimorphism, a point on which there are no data for ground sloths. Moreover, Paula Couto points out on a previous page that the smallest specimens referred to *M. r. casimbae* are aberrant and may not belong in *Megalocnus*. If two specimens of such highly doubtful identification are omitted, the ratios of lengths of the largest specimens placed in *M. r. rodens* to the smallest placed in *M. r. casimbae* are only 1.29 for the humerus and 1.22 for the femur. Such figures are usual in single populations of mammals and do not in themselves suggest any taxonomic distinction even at a subspecific level.

Matthew's figure of 300 per cent (i.e. of a largest/smallest ratio of 3.00) and his recognition of more than one species in *Megalocnus* were based on the comparison of very early juvenile lower jaws and teeth with corresponding parts of the largest adults. The distributions of available measurements of these specimens may be exemplified by those of M_3. That tooth provides the largest sample and was singled out in diagnosis of supposed species. Other distributions, although not always so clear in their smaller samples, are similar in pattern and consistent with conclusions based on M_3, so that there is no reason to use space for presentation and discussion of more than the one example.

The data for M_3 are graphically summarized in Fig. 5–1. There is a single cluster of numerous specimens, and there are three specimens that fall distinctly outside the cluster. The cluster appears to be a sample of a single population, with each of the two dimensions in an approximately normal distribution. The variation, as is further shown in Table 5–1, is such as commonly occurs in fully homogeneous samples of single populations of mammals. The estimated coefficient of variation, 6.7, is quite usual for single subspecies, even for teeth that are, unlike these, rooted and therefore nearly constant in size in a given individual. Specimens referred to both *M. r. rodens* and *M. r. casimbae* are included, but neither the statistical data nor the graph suggests any difference between them.

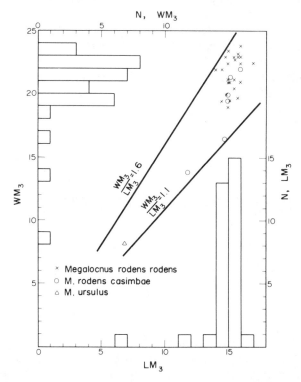

FIG. 5–1. Graphic presentation of separate and bivariate distributions of measurements of length and width of wear surfaces of third lower molars in samples of the Cuban ground sloth genus *Megalocnus*. Lengths in millimeters and corresponding frequencies in the sample are given in histograms for length, at bottom, with scale of frequencies to the right, and for width, at left, with scale of frequencies at the top. The bivariate positions of the individual specimens are shown in the body of the diagram, using the same scales as for the histograms. The previous identification of these probably conspecific specimens is indicated, and they are shown all to lie between the values 1.1 and 1.6 for the ratio of width to length.

TABLE 5–1

Statistical Data for Width of M₃ in All Surely Adult and Precisely
Measurable Specimens Referred to Megalocnus rodens rodens
(N, 21) and Megalocnus rodens casimbae *(N,3) Combined.*

Number of Specimens (N)	Mean ($\bar{x}$)	Standard Deviation (s)	Coefficient of Variation (V)	Observed Range (OR)	Ratio of Upper to Lower Limits of Observed Range
24	21.7 ± 0.3	1.45 ± 0.21	6.7 ± 1.0	19.0–23.8	1.25

The whole picture suggested by this graph (and by all the other metrical data at hand) is that of a sample mainly of adults but with a few juveniles, all from a single population. The teeth (also the jaws and other bones) grew rather rapidly in the juveniles and changed somewhat in proportions. In adults, after the bones ceased to grow, the teeth continued to grow longitudinally (vertically) but no longer changed their horizontal dimensions significantly.

It is impossible to prove beyond any question and by biometrical means alone that a sample is taxonomically homogeneous. The data cannot establish the negative conclusion that Matthew's taxa are invalid. Nevertheless it seems fairly clear, as Paula Couto also indicates in the preceding text, that there is not sufficiently sound evidence of their validity. Establishment of a statistical difference between local populations, such as Matthew suggested between his *M. e. rodens* and *M. r. casimbae*, would require larger samples of definitely and precisely recorded provenience than are now available. Establishment of a specific difference, as suggested in *M. rodens*, *M. ursulus*, and *M. junius*, would require discovery of specimens *of the same age* and demonstration that they are not conspecific.

Source

Appendix (pp. 51–54) in:
    MATTHEW, W.D., and C. DE PAULA COUTO (1959) The Cuban Edentates, *Bulletin of the American Museum of Natural History*, Vol. 117, pp. 1–56.

Range as a Zoological Character

Comment

The reasons for the proposal in a paper under this title are sufficiently given in the following excerpt from the paper itself. I add that although zoologists now are usually more sophisticated in biometry than they were in 1941, many of them do still seem to consider the observed range as a reasonable indication of the population range. I might otherwise seem to be flogging a dead horse in reprinting this attack on, and remedy for, that point of view.

Only the first part of the original publication, its pages 785 to top of 798, is here reprinted. Three paragraphs (on pages 786–787), discussing general approaches to the statistics of ranges, are unnecessary here and are omitted. In the later omitted part of the paper there was a discussion of estimating standard deviation and standard range from observed range and a table for that purpose, calculated by me from the data in Table 5–2, was given. As the calculation of standard range from standard deviation is preferable and calculation of standard deviation is so easy with present generally available means, the use of observed range in that way has become pointless. There were also a number of examples of applications of range concepts to particular samples of fossil mammals which also seem unnecessary for present purposes.

Text

Every study of fossil and living animals is necessarily aware of the existence of variation in his materials and is obliged to make some measurement or estimate of that variation. The most obvious measurement is provided by the difference between extreme values of the variate in question, for instance between the greatest and the least lengths of bones or shells in an available sample. This range of variation in a given series of specimens is what is actually seen by the student and ideas as to variation are intuitively based on it. It is a common assumption that this is really a measure of variation in the species to which the specimens belong, and in zoological and paleontological literature this is by far the most common figure given, in some form, if variation is considered at all. In fact the observed range in a sample is probably the least reliable and enlightening of all available measures of variation, a disadvantage still insufficiently appreciated by some workers but generally known and adequately discussed by those familiar with modern quantitative methods in zoology. It is unnecessary to belabor the point here. The meaning of the observed range is not well known and its use in the literature is usually wrong, but it does have a meaning and legitimate uses. The purpose of this note is to explain the meaning as simply as possible and to facilitate the uses.

The primary measurement of range in a given set of specimens is simple and obvious enough, but with rare exceptions no particular importance attaches to this measurement for its own sake. It is valuable only as an indication of the range, or variation, in a population of which the observations in hand are a sample. It may, for instance, be desirable to know what the range is in a given species, and then the observed range in a sample is useful only to the extent that it provides an estimate of the specific range. The literature is full of statements that the range in a species is some definite figure, but the figure given is always that of a sample, not of the species, and the two things thus confused are not only distinct in kind but also different in numerical value.

A variate of a species always does have a largest and a smallest real occurrence, so that the population range in this and analogous cases is something that has a real existence in nature. It is virtually never available for direct observation, however, because the chances of observing the one largest and the one smallest value are infinitesimal, and even if this should occur there would be no way of knowing that these were indeed the terminal members of the population. Moreover, the population range is not constant. It has a real value at any given instant, but it may fluctuate greatly. The largest individual of a species may die tomorrow or a smaller individual may be born. The most desirable datum on range would be its average value in a population under defined conditions and regardless of random changes.

Like the real population range and to still greater degree, the average or mean population range cannot be directly observed, but it is the value of which an estimate is desired. When such an estimate has been obtained, it is desirable to have some idea of how good the estimate is likely to be. It is also useful to know how much fluctuation in range may reasonably be expected in samples available for observation.

If obtaining and publishing an observed range from a given sample is supposed to give information about the species, or the population in a less specific sense, this carries the implicit assumption that the sample range is an estimate of the population range. It would be a valid estimate only if it tended to have the same value as the latter, aside from the random effects of chance in sampling. This is never true. The sample range is (for practical purposes) always smaller than the population range and it never tends to have the same value. With samples such as are usually available to zoologists and paleontologists, the sample range seldom tends to be over half as large as the population range. Moreover, the tendency is different in samples of different sizes so that the direct comparison of ranges derived from different numbers of observations is incorrect and usually grossly misleading, despite its widespread use.

It is superficially paradoxical but it is true that the range of a population has no direct bearing on the range of a sample drawn from that population. Conversely, the sample range does not directly give any information about the population range. The range in a sample really depends on these factors:

(1) chance deviation from the value toward which sample (not population) ranges tend under the given conditions;

(2) the size of the sample;

(3) the unit of measurement;

(4) the absolute value of the mean (not mean range) in the population;

(5) the dispersion of the population.

Samples of a given size from a given population do tend toward a fixed value, determined by the other factors, (2)–(5). This is the mean value for samples of that size and may be briefly designated as the mean sample range. If the frequency distribution of the variate in question is approximately normal, as is true of most zoological variates, the mean sample range is definable in terms of only two variables: total frequency of the sample (i.e. number of individual observations made) and standard deviation of the population. If the distribution is not normal (in a statistical sense or an approximation of a symmetrical binomial), the problem becomes quite different and its discussion is outside the scope of this paper. The simple solution for normal distributions is made possible by the fact that the standard deviation involves all the factors numbered (3) to (5) above. It is in terms of the unit of measurement and is a measure of dispersion that tends to be proportional to the mean, for like amounts of variability. The standard deviation of the population, like its range, cannot be obtained by direct observation, but it can be estimated from a sample by statistical methods already in wide use and unnecessary to explain in this paper. It is thus possible to work backwards and to use the sample size, which is given, and the estimated standard deviation, which can be calculated, to obtain an estimate of population range. A particular sort of estimate, proposed in this paper, may be designated the standard range.

Evidently the word "range" is used for several different things and its application may be made precise by the following terms and definitions:

Observed range: a measured sample range. (Symbol: *OR*.)

Mean sample range: value that ranges in samples of a given size from a defined population tend to approximate.

Expected range: an estimate of mean sample range for samples of any one stated size, as calculated by procedures given later in this paper. (Symbol: *ER*.)

Standard Range: an estimate of mean sample range (i.e. the expected range) for samples of a constant, standard size, set at 1000 in this paper. This is the same as an estimate of population range on the assumption or convention that the population frequency is 1000. (Symbol: *SR*.) Two methods of estimation will be proposed: standard range from standard deviation, *SR* (*SD*), and standard range from observed range, *SR* (*OR*).

Population range: the real range in a finite population. It has fluctuating individual values and also, for any given size of population with definite specifications, a constant mean value. This, the mean population range, is the same as mean sample range for the same abundance or frequency.

The mean sample range can never and the population range can only very exceptionally be obtained by direct observation, but the terms are necessary for clarity of discussion and for maintaining the distinction between these unobtainable values and the estimates of them.

It has been stated that the mean sample range depends on sample size and population standard deviation. It is therefore possible to express the mean sample range for any given size of sample as the standard deviation multiplied by a constant. The relationship is intricate and the work of solving the equation for determination of this constant is beyond the patience, if not the ability, of most of us who are not mathematicians, but the work has already been done by Tippett for all sizes of samples from 2 to 1000. His results are summarized in the accompanying simplified and condensed table (Table 5–2, Column (A).

If the standard deviation of the population were known, the mean sample range for any size of sample could be obtained from this table by the product of the factor in Column (A) times the standard deviation. For instance, if the standard deviation were 10 mm, the mean sample range for samples of four specimens each would be 20.6 mm, for samples of eleven specimens each 31.7 mm, etc.

In fact the population standard deviation cannot (unless very exceptionally) be directly obtained, but an estimate of it can be made from a sample in various ways, most reliably by actual calculation of the standard deviation of the sample. Since standard deviation, unlike observed range, does not tend to vary with size of sample or population, this sample standard deviation requires no conversion for use as an estimate of the population value. Given such an estimate of the standard deviation, a corresponding estimate of mean sample range can be calculated from the table and this is the expected range or *ER*. For instance, if a sample of 12 specimens has standard deviation 8.1 mm, the *ER* for samples of that size is $8.1 \times 3.258 = 26.4$ mm. The *ER* for samples of 100 specimens each from the same population is $8.1 \times 5.015 = 40.6$ mm and for samples of 1000 each, $8.1 \times 6.483 = 52.5$ mm. Note that in a sample of 12 specimens, the range tends to be only half as large as for a sample of 1000 drawn from exactly the same population— evidence of how misleading is much of the use of observed ranges.

Of course any one sample will be unlikely to have exactly the expected range. This is an estimate of mean sample range and individual observed ranges will fluctuate around this mean. It is useful and necessary to have some idea as to how much any one observed

TABLE 5–2

*Conversion data for calculation of expected ranges
and related figures*

In terms of standard deviation:

Size of Sample	(A) Mean sample range	(B) Standard error of range	(C) 1% points of range distribution
2	1.284	0.853	0.02–3.64
3	1.693	0.888	0.22–4.10
4	2.059	0.880	0.47–4.38
5	2.326	0.864	0.70–4.59
6	2.534	0.848	0.89–4.74
7	2.704	0.833	1.07–4.87
8	2.847	0.820	1.22–4.98
9	2.970	0.808	1.36–5.07
10	3.078	0.797	1.48–5.15
11	3.173	0.787	1.59–5.22
12	3.258	0.778	1.69–5.28
13	3.336	0.770	1.78–5.34
14	3.407	0.762	1.86–5.39
15	3.472	0.755	1.94–5.44
16	3.532	0.749	2.01–5.49
17	3.588	0.743	2.08–5.53
18	3.640	0.738	2.14–5.57
19	3.689	0.733	2.20–5.51
20	3.735	0.729	2.25–5.64
22	3.819	0.720	2.36–5.71
24	3.895	0.712	2.45–5.76
26	3.964	0.705	2.53–5.82
28	4.027	0.699	2.61–5.87
30	4.086	0.693	2.69–5.91
35	4.213	0.681	2.84–6.01
40	4.322	0.670	2.97–6.09
45	4.415	0.660	3.09–6.16
50	4.498	0.652	3.19–6.23
55	4.572	0.645	3.28–6.29
60	4.639	0.639	3.36–6.34
65	4.699	0.633	3.43–6.38
70	4.755	0.628	3.50–6.43
75	4.806	0.624	3.56–6.47
80	4.854	0.619	3.62–6.50
85	4.898	0.615	3.62–6.50
90	4.939	0.612	3.72–6.57
95	4.978	0.608	3.77–6.60
100	5.015	0.605	3.81–6.63

Continuation of Column (A) for larger samples

Size of sample	Mean sample range	Size of sample	Mean sample range
150	5.298	500	6.073
200	5.492	550	6.131
250	5.638	600	6.183
300	5.756	700	6.275
350	5.853	800	6.354
400	5.936	900	6.422
450	6.009	1000	6.48287

Note: The data for Column (A) are from L.H.C. Tippett, those for Columns (B) and (C) from E.S. Pearson.

range is likely to differ from the expected range. In analogous cases, for instance, in considering the scatter of means of a number of samples around their common mean (the mean of the sample means), this is estimated by the use of a standard error. Tippett and E.S. Pearson have calculated standard errors of ranges corresponding with various sizes of samples and their factors are given in Table 5–2, Column (B). For example, if the standard deviation is estimated as 8.1 mm, the standard error of range for samples of 12 each will be $8.1 \times .778 = 6.3$ mm and the *ER* is 26.4 ± 6.3 (or, giving only significant figures, 26 ± 6). In any one sample, the value of a given statistic, such as range, is unlikely to differ from the mean (or expected) value by more than three times the standard error, or in this case by more than about 18mm. Since the expected value for samples of 12 in this case is 26 mm, as previously found, it can therefore be inferred that no sample of this size from this population is likely to have a range of more than about 44 mm or less than about 8 mm.

This procedure is based on the assumption that the values in question, ranges in this case, tend to be symmetrically distributed around their mean, or expected value. This is not exactly true of ranges and although the procedure is usually safe enough as an approximation, it is not strictly correct theoretically. The theoretical probabilities based on an asymmetrical distribution have been calculated by E.S. Pearson and Column (C) of Table 5–2, giving 1 per cent points, is summarized and simplified from his more elaborate data. Like Columns (A) and (B), the entries are factors to be multiplied by the standard deviation in order to obtain the values sought. A "1 percent point" is a value such that there is only one chance in a hundred that it will be exceeded (for the upper point) or not reached (for the lower point) by a sample drawn at random—a chance small enough to be considered negligible in most practical work.

For samples of 12, the tabled 1 per cent factors are 1.69 and 5.28. For standard deviation 8.1 mm, for instance, as before, this makes the 1 per cent range points $8.1 \times 1.69 = 13.7$ mm and $8.1 \times 5.28 = 42.8$ mm. In other words it is very unlikely (the odds are about 99 to 1 against) that a sample of 12 from this population will have an observed range of less than about 14 or more than about 43 mm, limits closer and more useful than those derived from the standard error of range. For samples of 100 from the same population the corresponding limits are 31 and 53, and it is interesting to note that in such samples of identical origin the observed range in 100 specimens may be nearly 4 times as great as in a sample of 12 specimens although, of course, the difference will usually be less.

The estimation of sample ranges and their limits, just discussed, has many zoological applications, but in this field it is more commonly desired to estimate from a sample what the population range is likely to be.

It has been shown that the mean sample range for any given size of sample can be estimated if an estimate of the standard deviation is available. The suggestion may at once be made that the population range could be estimated in the same way by calculating the expected range for a sample equal in size to the population, or, considering the population as potentially infinite, for a sample of infinite size.

A population of infinite size would theoretically have infinite limits and infinite range. In fact the concept or reality of range is absent in considering the mathematically perfect and infinite normal curve, or related representations of distributions. The theory and reality of ranges depend on the fact that real populations are finite, like samples, and the mean population range can be estimated only on the basis of a population of some given size.

That the population range, like the sample range, tends to vary with population size or abundance is a fact that has seldom been taken into account in zoological practice, but that must be considered if the term population range or equivalent expressions such as "range of variation in species" are to have any real meaning. A small population tends to have a smaller range than a large population even though their biological variability is the same or indeed even though they are recently and adventitiously separated groups of exactly the same race. This is another reason why the ordinary use of ranges in zoology has so little meaning and validity. If, however, population abundance were taken into account, range could be used in a scientifically sound way.

Use of an estimate of the real abundance of the population is necessary for any attempt to estimate real range, but with some possible exceptions the procedure is not practical. There seldom is any way of making a good estimate of population abundance. Estimates of range are usually wanted for purposes of comparison and the real population abundance may invalidate comparisons and be extraneous to the problem. Most species are so abundant that the factors for estimating their mean population ranges have not been calculated.

These difficulties are almost insuperable and they suggest the expedient of taking some standard of abundance and estimating what the mean population range would tend to be if the population had that abundance, or when it had it. This is an artifice, of course, but it is a valid and necessary convention. Moreover, the mean range for one conventional abundance is a fixed character of the population (and a statistical constant) while real population range is a variable. The use of a very large standard abundance, say one million, would have the advantages of being in a region where large differences in abundance make very little difference in mean range and of probably having the same order as the real abundance of many animal species. It has, however, the crippling disadvantage of being far beyond the region for which the necessary mathematical data are now available and the fatal disadvantage of being equally far beyond the region where any empirical check on the biological applicability of the theory has been made or is reasonably possible. Empirical checks are possible and have been made for smaller zoological samples and they do follow the statistical theory.

For these reasons I propose adopting 1000 as standard abundance. This number is small enough that virtually all natural groups do reach or exceed it and that the statistical theory can be extended to it with confidence. It is large enough to be in a region where the mean range is beginning to change slowly with changes in abundance, to give at least the right order of magnitude for a range of any probable abundance, and usually to be as close to the real population range as the span of the limits of sampling error.

Given an estimate of standard deviation, the mean population range for population abundance of 1000 can be estimated by use of the appropriate factor from Table 5–2, which is 6.48 (more exactly 6.48287). The standard deviation times this factor gives what is here defined as *standard range from standard deviation or SR (SD)*. Its great advantage over the observed range in practical work is that it tends to have the same value when based on samples from the same population, regardless of the sizes of the samples.

It cannot be assumed that the standard deviation calculated from a given sample is exactly that of the population. Two samples from the same population are not likely to have exactly the same standard deviation, and hence the same *SR (SD)* although these will tend toward the same central value. It is useful to estimate the highest and lowest *SR (SD)* values that would be likely to arise from other samples drawn from the same population as the sample in hand.

One per cent points of the standard deviation occur at ± 2.576 times its standard error. That is, there is only about one chance in a hundred that samples from a given population will differ by more than this amount from the population value. From this relationship it is possible to estimate approximately what the population standard deviation would be if an observed standard deviation happens to be at the minimum and if it happens to be at the maximum probable value relative to the population value. Hence maximum and minimum values of the *SR (SD)* likely to arise from samples drawn from the same population can be calculated. Using the previous data, the work is as follows:

Sample size, 10
Standard deviation, 0.49   } Calculated from sample.
Standard error of standard deviation, 0.11

$0.11 \times 2.576 = 0.28$
$0.49 + 0.280 = 0.77$,   probable maximum value of population standard deviation.
$0.49 - 0.280 = 0.21$,   probable minimum value of population standard deviation.
$0.77 \times 6.483 = 5.00$,   probable maximum value of mean range in samples of 1000.
$0.21 \times 6.483 = 1.30$,   probable minimum value of mean range in samples of 1000.

Another theoretical range already in some use is six times the standard deviation. The methods suggested in this paper are preferable but the previous method is equally valid (given its premises) and can be related to the same logical considerations. Although this has not previously been noted, the use of six times the standard deviation is equivalent to estimating the mean range for a population of about 450 individuals, as may be deduced from Table 5–2. (The exact figure, from Tippett's more elaborate tables, is between 442 and 443.) There is no reason why this should not be taken as standard population abundance, but 1000 is preferable when the procedure is put definitely on this basis.

Source

(1941) Range as a Zoological Character, *American Journal of Science*, Vol. 239, pp. 785–804

Another reference

Range, including the standard range, is discussed more briefly on pages 139–142 of SIMPSON, ROE, and LEWONTIN (1960)

Standardization of Normal Frequency Distributions

Comment

If range is to be discussed at all, the necessity for replacing observed range by some sort of standardized range based on standard deviation and on a convention as to sample or population size can hardly be disputed. Practical procedures, as discussed in the previous selection in this book, depend on the assumption that the distribution of the variate under study is approximately normal. That is nearly enough correct for practical purposes in most, but not all, variates usually studied in systematic zoology. At that point in our studies it occurred to Dr Roe and me that it would be useful if an observed sample distribution could be directly compared with the normal curve and with similar distributions in other samples or in other variates of the same sample. That would evidently require some graphic method that would show the distributions and a normal curve independent from the absolute dimensions of the variates studied. That was the line of thought that led us to write the paper an extract from which is here presented.

In the original paper we placed some emphasis on ease of calculation and included methods of working from observed range. That is not now really important, and it can now be assumed that the standard deviation will be estimated from the sample and a standard range calculated from that for an assumed sample size, 1000 in our work. In the following excerpt I have therefore eliminated some introductory remarks and discussion of the possible use of the observed range. I have also omitted discussion of the comparison of the sample histograms with the normal curve, some examples of use of the method, and some tables meant to simplify calculations. From the examples I have, however, here included in Fig. 5–3 one as displayed on the standard chart.

In the text the standard range was specified as SR (SD), for standard range from standard deviation, but that is here emended to SR alone because the standard range should always be derived from the standard deviation. Following more recent usage the mean, here designated M, should now be given as $\bar{x}$ and the sample estimate of standard deviation should be given as s, reserving sigma for the (unknown but estimated) population value.

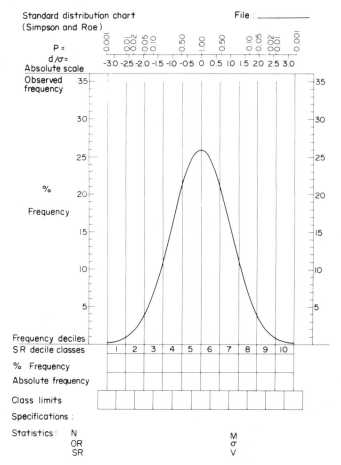

Standard distribution chart
(Simpson and Roe)   File : _____

P =
d /σ=
Absolute scale   -3.0 -2.5 -2.0 -1.5 -1.0 -0.5 0 0.5 1.0 1.5 2.0 2.5 3.0

Observed   35   35
frequency

              30   30

              25   25

%
              20   20
Frequency

              15   15

              10   10

               5   5

Frequency deciles
S R decile classes | 1 | 2 | 3 | 4 | 5 | 6 | 7 | 8 | 9 | 10 |

% Frequency

Absolute frequency

Class limits

Specifications :

Statistics: N   M
               OR   σ
               SR   V

FIG. 5–2. Standard chart for work-sheets and filing, as described in the text.

Text

The principal aims of the present method are:

(1) to provide a convenient standard from for recording and publishing data on frequency distributions;

(2) to bring together concisely all the data that are likely to be needed for further uses;

(3) to specify a limited but widely useful armament of statistics;

(4) to provide a means of graphic presentation in which differences due to size of sample, units of measurement, mean value and other factors extraneous from the given point of view are systematized and minimized or eliminated;

(5) to combine with this a standard normal curve that will be roughly fitted to the graphic distribution;

(6) to provide simple graphic and tabular means for estimating the statistical significance of deviations, and making related tests;

(7) to facilitate various other special uses such as the scoring of test results.

These aims are furthered by the provision of a printed chart or record blank and by the explanations and methods of the following text.

The theoretical infinite normal distribution is unlimited and so has infinite range, but real populations and samples of them are finite in frequency and in range. Range of finite groups is variable but has a constant central tendency or mean value. This constant depends on the dispersion of the population (as best measured by the standard deviation) and on the size of the sample or real population. Since the range of samples from the same population has quite different mean values for samples of different sizes and has no constant relationship to the population range, the observed or sample range has no meaning for population studies unless related to sample frequency and cannot be compared for different samples unless reduced to some one standard frequency.

For various practical reasons, the standard frequency chosen is 1000. The standard range is an estimate, derived from a given sample, of the mean value of the range in samples or finite populations of 1000. For any given sample size, the mean range divided by population standard deviation is constant. For the standard frequency 1000, this ratio is 6.48. Therefore, if the population standard deviation has been estimated from the sample, multiplying this by 6.48 gives an estimate of standard range, SR.

Sample distributions represent population distributions in an imperfect but systematically variable way, and statistics derived from samples are used as estimates of the corresponding characters of the populations as a whole. Judgment as to these sample-population relationships and their use for estimates of probability of deviations, and the like, are customarily derived from various statistical tables or by diverse graphic methods that approximate the observed data to idealized normal (or sometimes other) curve data. Graphic approximation has many advantages in conciseness, ease of visualization and frequently in ease or elimination of much calculation.

The basic principle of such graphic methods is to divide the observed frequency distribution into a convenient number of defined classes and to compare the results with approximately fitted ideal distributions and statistics. The classes are decile classes if 10 in number and percentile classes if 100 in number. They may be classes of equal frequency, and then are frequency decile and percentile classes or ordinate decile and percentile classes, that is, a frequency decile class contains one-tenth of the total frequency of the distribution. The variable to be treated and compared is then the span or range of values covered by each decile class (or percentile class). Simple graphic use of ordinate deciles necessitates plotting of the cumulative distribution, which is disadvantageous because such a plot does not easily convey the real form and character of the distribution.

Instead of having equal frequency, the classes may be defined as having equal range of values of the variate, that is, they are equal subdivisions of the abscissa of the usual graph of a frequency distribution, and deciles so defined are abscissal deciles. Then the variable for representation, study and comparison is the frequency included in each abscissal decile class. This is the method used in this paper.

Full specification requires a statement not only that abscissal deciles are used but also whether the frequencies are absolute or relative and how the span to be divided into ten equal parts is determined. Use of absolute frequencies, those observed in the given sample, is of course a customary ordinary method of representation of a distribution, for instance, in a histogram, but it will not serve the purposes here in view because its use means that distributions with identical statistics may appear very different and because every sample will probably approximate a different form of the normal distribution, requiring separate tabling or calculation. These disadvantages are minimized if relative frequencies are used in place of absolute. Percentages of total sample frequency will be used as a convenient and generally understood form of relative frequency.

Ordinate deciles are parameters, or constants, of the normal curve because the curve has a determinate, finite area (proportional to frequency), and this can be divided into ten equal parts. Abscissal deciles are not necessarily parameters of that curve (or of other ideal distributions), because the range of the curve is theoretically infinite and cannot be divided into finite parts. Therefore the same artifice must be employed as for the standardization of range, the curve must be taken as terminating at some specified point and as having finite range. Divisions of the observed range into ten equal parts, giving what may be called observed range or *OR* deciles, is unsuitable because the systematic variation of *OR* with N defeats the purpose of comparing samples of different sizes and involves other decided disadvantages.

From the background of standard range, *SR*, methods, it is clear that the *SR* deciles will fulfill the requirements. *SR* is a constant of the normal distribution (does not systematically vary with N or other extraneous sample characters), it can easily be estimated more or less closely from samples of any size greater than two, and it almost always includes all the observed frequencies. *SR* is, then, the total abscissal distance to be used for our standard distributions. The length of each *SR* decile is $SR/10$. The lower limit of the distribution is $M - SR/2$ and the upper limit is $M + SR/2$, in which M is an estimate of the population mean from the sample. The sample mean thus falls in the middle of the diagram for all samples and coincides with the mean of the comparative normal curve. Successive decile limits are obtained by successive addition of $SR/10$ starting at $M - SR/2$, or successive subtraction of $SR/10$ starting at $M + SR/2$.

With the *SR* decile class limits determined in this simple way, the observed frequencies are entered in the classes and are converted into relative (%) frequencies by dividing by N (sample frequency) and multiplying by 100. This, the relative *SR* decile distribution, is the standard distribution of the present method. It may now be graphically portrayed as a histogram and compared with a standard normal curve, appropriate (with certain limitations discussed below) for any roughly normal observed distribution treated in this way and drawn once and for all on the standard chart in Fig. 5–2.

The frequency deciles for the standard normal curve can also be tabulated, as inserted on the chart. These completely normal frequency deciles will not correspond exactly with those calculated from the sample but will approximate the latter for samples of fair to large size. More important, these charted frequency deciles, obtained without any calculation specifically for this purpose, are better estimates of population frequency deciles (if the population is normal or nearly so) than can be obtained by relatively laborious calculation directly from the observed values.

At the top of the chart there is a scale of deviations in terms of standard deviations, d/σ, and of corresponding sampling probabilities, P. P gives as a fraction of 1.00 the probability of random drawing from the normal population of an observation with equal or greater deviation from the mean. The d/σ and P values are those of the standard normal curve. If *SR* has been derived from standard deviation, giving *SR (SD)*, the d/σ and P of the graph and tables will be identical with estimated values calculated from

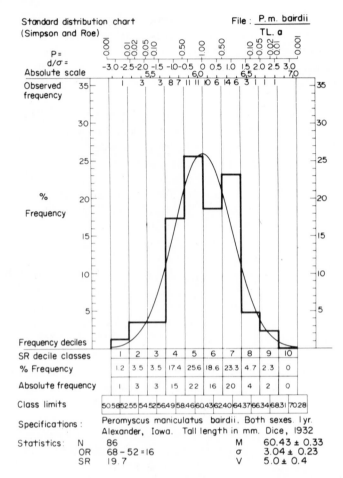

Standard distribution chart   File : P. m. bairdii
(Simpson and Roe)   TL. a

FIG. 5–3. The standard chart completed for one dimension of a subspecies of Recent white-footed mice.

the sample. The charted values are usually sufficiently good estimates for samples of moderate or large size, $N = 15$ or more. When N is less than 15, it is usually preferable to obtain P from the distribution of t.

Source

SIMPSON, G.G., and A. ROE, 1942. A standard Frequency Distribution Method, *American Museum Novitates*, No. 1190, pp. 1–19.

6

Biogeography

THE GEOGRAPHIC distribution of organisms is of great importance for many aspects of evolutionary biology. It is in itself a complex subject with numerous different approaches and methods. Ecological biogeography is concerned with the environmental conditions under which single groups (taxa) or communities of organisms occur. Historical biogeography has to do with where a taxon or a community has occurred in the course of its geological span and how it came to be where it was and, if still living, now is. Another approach, which was dominant in the beginnings of biogeography as a science, involves the recognition, definition, and designation of regional faunas and floras on scales varying from a few square meters to whole continents or more. (One major named recent zoogeographic region, the Palearctic, occupies all of Europe and large parts of Asia and Africa; another regional designation, Holarctic, adds most of North America to that.) Such studies necessarily involve comparisons of floras and faunas as wholes, and those sometimes difficult comparisons are now often quantified in various ways. Another basically descriptive and now quantified biogeographic procedure studies the regional densities in numbers of species or other taxa. Especially when dealing with ancient faunas and floras there are also sampling problems: to what extent are the known fossils representative of regional faunas or floras as a whole?

Most of those different approaches and topics are discussed and exemplified in the present section of this book. Ecological biogeography is not, except to the extent that it is a determining factor in species density. Ecology is touched on in other parts of this book. A fairly recent quantitative and modeling approach to one aspect of historical biogeography, based in the first instance on island faunas, is not here exemplified, but I give a reference to its fundamental document at the end of this introduction.

A recent controversy should be mentioned briefly here, if only because I have not discussed it in any of my rather numerous previous publications on biogeography. The reason for my having avoided it is that I consider it as absurd and not worthy of the attention it has received in some restricted circles. This is the so-called vicariance biogeography. Its main point is that new taxa do not usually arise in a limited area and then disperse from there but more often arise by separation of the population of a taxon, originally covering the whole area involved, into more local populations which then evolve into distinct, "vicariant" taxa. The absurdity is that the exponents of this concept go on to say that in the great majority of cases related taxa do arise allopatrically, that is in different and limited areas; in fact they use "vicariance" as a synonym of "allopatry." They then note that a taxon originating allopatrically frequently spreads or disperses from its area of origin. In the main they are thus using a different jargon to express exactly what the biogeographers they attack, sometimes viciously, had been saying all along. The absurdity extends to some of the other concepts and methods advanced by the disciples of "vicariance." (It should be mentioned that the term "vicars" has long been used by biogeographers not proponents of "vicariance biogeography" to refer to related species living in similar ecological conditions in different areas.)

References

COX, C.B., I.N. HEALEY, and P.D. MOORE (1973) *Biogeography, An Ecological and Evolutionary Approach*, John Wiley and Sons, New York. (Most books on biogeography are long on description and details, short on synthesis and theory; this one comes close to giving an introduction to the whole field.)
MACARTHUR, R.H., and E.O. WILSON (1967) *The Theory of Island Biogeography*, Princeton University Press, Princeton.

Approach to a Sampling Problem

Comment

The following paper, reprinted with little change, was an early approach toward judgment of the probability that the known fossils of a limited region and facies indicate the general nature of a once broader, especially continental, fauna.

The sampling of the Eocene of North America and the pre-Pliocene of South America is now very much better, both in local completion and in areal extent, than it was in the 1930s as indicated in the first paragraph of this paper. These significant, welcome increases in knowledge nevertheless support the contention that even in the 1930s the described collections did give a fair overall, although not detailed, idea of the continental faunas at those times.

Text

The notorious incompleteness of the paleontological record is due not only to gaps in time distribution of known fossils, but also, and at present in still greater degree, to the relatively local character of most of the faunas known. It is often necessary to judge the mammalian fauna of a whole continent or major land area from fossils found only at one or a few localities. For instance, the Eocene land life of North America is known almost exclusively from discoveries in a few basins scattered through the Rocky Mountain area. It is usual to speak of the Eocene fauna of North America, but in fact for any one part of the Eocene only a relatively very small and localized geographic group of that general fauna is known. Other instances are even more striking, as in South America, where almost all known pre-Pliocene mammals are from the relatively limited and marginal areas of Patagonia. It is obvious that such data cannot give a complete picture of the fauna of a whole landmass.

In the endeavor to visualize these relationships and to give some concrete basis for their evaluation, detailed comparison has been made between the Recent mammalian faunas of Florida and of New Mexico. The results, here set forth in outline, clearly are no more than an approach to the general problem, which requires very many more data for any full consideration or definitive solution. They do, however, suggest a means of approach which does not appear to have been utilized in this way previously, and they give some definite basis, however inadequate as yet, for judgment in an important matter hitherto generally ignored or treated by guesswork.

Florida has a land area of 54,861 square miles and a water area of 3,805 square miles. It is a marginal region and is almost surrounded by the sea. Physically it is low, with little relief or physiographic variety (maximum elevation, 325 feet; minimum, sea level; average 100 feet). The climate is oceanic and is warm and humid. Faunally, parts of the Lower Austral and Tropical zones are included. The mainland lies between 25°07′ and 31° north latitude.

New Mexico has a land area of 122,503 square miles and a water area of 131 square miles. It is central in position and far from the continental borders. The physiography is extremely varied, including high mountains, plateaus, plains, desert basins, and low alluvial valley bottoms (maximum elevation, 13,306 feet; minimum, 2,876; average, 5,700 feet). The climate is markedly continental, with hot summers and cold winters, and in general is semi-arid. It includes parts or representatives of all fauna zones from Hudsonian to Lower Sonoran. It lies between 31°20′ and 37° north latitude.

The closest points of the two states are over 900 miles apart, and their centers are separated by about 1,450 miles. They are markedly different in position, environmental and physiographic variety, climate, and faunal zones. It is probably impossible to duplicate exactly in either an environment found in the other. They have in common the facts that both are distinctly parts of the North American continent of long standing, from the point of view of Recent mammals, and that both lie in the southern part of the Temperate Zone.

The primary object of the comparison is to see what could safely be inferred of the fauna of the other if the mammalian fauna of only one of these two states were known. Hence some judgment is possible as to the limits of safe inference regarding the regional fauna in cases of fossil faunas known only from one part of a large landmass.

Lists of recent groups of land mammals of the two states have been compiled, and while no two students would agree as to exact numbers, the relative proportions are not greatly dependent on any personal factor. The following figures represent the number of orders, families, genera, and species (subspecies being excluded) of Recent terrestrial mammals, exclusive of man and his importations, native to each state, and the numbers common to both.

	New Mexico	Both States	Florida	Total
Orders	6	5	5	6
Families	19	12	13	20
Genera	54	18	27	63
Species	133	6	33	160

The data for each state are more graphically comparable in Fig. 6–1. The immediately striking fact is the greater variety of the New Mexican fauna, with four times as many species as Florida. The apparent reasons for this, in probable order of importance, are the greater environmental diversity of New Mexico, its larger area, and its more central position.

A second noteworthy fact is the markedly different ratios for different taxonomic

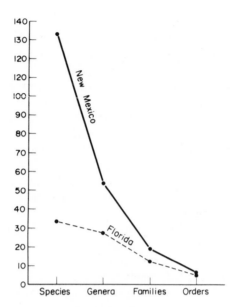

FIG. 6–1. Graphic comparison of numbers of species, genera, families, and orders of recent, indigenous land mammals in Florida and in New Mexico.

groups. New Mexico has four times as many species as Florida, but only twice as many genera and half again as many families, while the number of orders is nearly the same. The two curves of the diagram consequently converge to the right. If comparable curves are plotted for areas still larger or still more varied than New Mexico, they rise more and more steeply to the left, but the convergence to the right remains.

As regards this example, then, if we knew all the mammals of one state, we could infer almost exactly how many orders occur in the other, and fairly well how many families, but might be badly mistaken regarding the lesser units.

More important than the number of units is their character. Somewhat contrary to expectations, the example suggests that in similar cases and within certain broad limits, the character of the fauna of one region can be inferred from that of the other. These data are graphically shown in Fig. 6–2, in which the elements common to both states are contrasted with those peculiar to one. Another form of presentation is that of Table 6–1.

TABLE 6–1

	Common to both states	Peculiar to one or the other	Florida fauna occurring also in New Mexico	New Mexico fauna occurring also in Florida
Orders	83%	17%	100%	83%
Families	60	40	92	63
Genera	29	71	67	33
Species	4	96	18	5

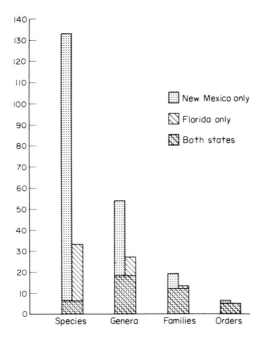

Fig. 6–2. Histograms of numbers of species, genera, families, and orders of recent, indigenous land mammals present in both Florida and New Mexico and in one of those states but not the other.

On this showing, knowledge of either fauna will give a fair idea of the general composition of the other, that is, of its orders and families. The possibility is also enhanced beyond the point indicated by the figures, and especially as regards smaller groups, by the fact that many mammals of each fauna have relatives in the other which are not identical but which are so similar that knowledge of one is applicable in a general way to the other.

Another striking fact, although one that might be assumed *a priori*, is that the larger fauna gives a better basis for inference as regards the smaller than does the smaller for the larger. New Mexico, with its highly varied fauna, gives a reasonably accurate picture of the mammalian life of at least the central part of North America. If we knew its fauna and not that of Florida, we would nevertheless be familiar with all the orders, almost all the families, and two-thirds of the genera that do occur in Florida. Even beyond this, although for instance, we would not know *Neofiber* or *Geomys*, we would have the closely related and ecologically similar genera *Ondatra* and *Thomomys*. There would in fact be only one Floridian genus, *Scalopus*, which was radically unlike any known from New Mexico.

Applied to paleontological data, the accuracy of such inferences depends on numerous factors which can be approximately evaluated, although any close approach to statistical accuracy would necessitate the accumulation and interpretation of many more data than have yet been studied, and would even then be an extremely complex and difficult problem. There is, however, no reason to believe it insoluble. The main factors requiring such consideration are probably the following:

(1) *The area represented by the collection.* This can be calculated, at least approximately, by the actual area covered by the geological formation, or by the probable size of the drainage area from which its fossils could have been derived. Accurate evaluation of this factor demands statistical data on the relationship of area and number of species under similar Recent conditions.

(2) *Environmental variety.* This can be estimated from the inferred physiographic conditions when the formation was being deposited, or perhaps more accurately from the nature of the fauna. A fauna which includes many cases of related but truely distinct contemporaneous species surely indicates environmental variety, for in general only one such species can live in one local environment, in one region, at one time (Cabrera's law of "ecological incompatibility"). Furthermore paleobiological inference may also indicate environmental variety, by indicating variety, or lack of variety, in the probable habits of the animals found.

(3) *Geographic position.* It can be reasonably established on paleogeographic grounds and sometimes in part from the fauna itself, whether the region from which a given fossil fauna is derived was at the time of that fauna part of the large landmass under consideration, and approximately what position it occupied on that landmass. Presumably the more central the area, the more varied its fauna is likely to be and the better it will represent the continental fauna as a whole. The influence of broad climatic zones also needs evaluation. Presumably a large fauna would include nearly 100 per cent of the important groups of mammals present on the same continent in the same climatic zone, and in passing to other climatic zones the percentage would be progressively lower. Within a reasonable margin of error, this relationship could probably be established mathematically. In the case of Tertiary faunas the factor would be less important than with Recent faunas, for the climatic zones were evidently much less marked during most of the Tertiary than they are at present.

(4) *Completeness of single faunas.* Accuracy of broad inferences must be directly proportional to the extent of knowledge of the fauna on which the inference is based. It is probably impossible in any case to obtain fossil representatives of 100 per cent of the groups of land mammals that actually lived in a given area. The percentage of representation can be roughly estimated from the size of the collection made, the percentage of new species added by new collections, the relation of the number of species known to the number which probably were present under the given conditions, the presence or absence of representation of given groups known to have occurred both before and afterwards in the same region, and the ecological character of the fauna, that is, whether all environments probably present are represented, whether both herbivores and carnivores are present in normal proportions, whether both large and small mammals are present, etc.

It seems probable that a large and varied Tertiary fauna which may be judged representative on these criteria, even if it is from a single and fairly local geological formation, gives an excellent representation of the mammalian orders present on the whole continent, a good idea of the families, and will probably even include most of the genera.

As an example, the mammals of the Bridger [middle Eocene of the Rocky Mountain region] fauna probably ranged over an area comparable to that of Florida. The position of their area was very central on the Eocene continent. The local environments represented are clearly numerous. Knowledge of the fauna is fairly complete. It follows that this fauna probably well represents the middle Eocene life of all of North America.

As an example of the sort of broader and more important inferences dependent on such data, the well known and apparently representative middle Paleocene faunas contain no possible ancestors of several Eocene orders and lesser groups. From the considerations here given, it therefore seems highly improbable that many of the typically Eocene groups were then present in North America, and failure to find them in the Paleocene cannot be entirely imputed to the localized nature of the known deposits of middle Paleocene age.

It seems probable that the paleontological data now in hand provide a better basis for studies of mammalian faunal succession and geographic distribution than is generally granted.

Source

(1936) Data on the Relationships of Local and Continental Mammalian Faunas, *Journal of Paleontology*, Vol. 10, pp. 410–414.

A History of an Intercontinental Faunal Resemblance

Comment

The following text is part of a long paper reporting a comprehensive study of the relationships between North American and Eurasian faunas of land mammals throughout the Age of Mammals (Cenozoic). The first part gave the detailed evidence for faunal interchange or extension during that span, with results much more briefly summarized on pages 206–209 of the present book. The second part of the paper, here excerpted with some omissions, discusses the history in terms of faunal resemblance.

In this paper faunal resemblance is quantified in terms of taxa at the generic and familial levels by the formula $100C/N_1$, in which C represents the number of taxa known to be common to both of the faunas compared and N_1 represents the total number of taxa known in the smaller of the two faunas. Although this coefficient is simple and would seem to be obvious, as far as I know it had not been proposed before I published it in 1943. The later (1947) application in the paper here excerpted provides more extensive examples and gives an analysis of the most important of them. The coefficient has since been widely used in biogeography and is sometimes called "Simpson's coefficient of faunal resemblance." In order to make its designation simpler and more explicit Flessa has recently proposed that the units of measurement (percentages) of the coefficient be called "simgens" when based on genera and "simfams" when based on families. The behavior and applications of this coefficient and its comparison with some others is more fully, although succinctly, discussed in the next excerpt reprinted in this book.

The geographic relationships of Europe, Asia, and North America are now better understood than they were in 1947, although even now not fully known. It is fairly clear that in the Paleocene and early Eocene what are now the separate continents Europe and North America were regions on what was essentially a single continent. (That was probably also true of eastern North America and Europe in the late Cretaceous, but the faunal evidence for that time is still inadequate.) Thus the high early Eocene resemblance, now known to have existed also in the late Paleocene, is explained by the faunas found on the now separate continents then having been only regional variants of the same continental fauna. It is now clearer that the early Cenozoic faunas of Europe and Asia were less similar than they later became, and this is explicable by the presence of a marine barrier between them: the Turgai strait and the Obik sea which was epicontinental, a shallow sea that existed on a single tectonic plate rather than between two plates.

There is also evidence of some intermigration in the late Cretaceous and early Cenozoic between Asia and western North America, which until about mid-Paleocene time was partly isolated from eastern North America by an epicontinental interior sea. That early Asian–North American interchange was limited in extent and may well have occurred by waif or sweepstakes dispersal across a somewhat passable sea barrier and not by a dry land connection. After the early Eocene it is now fairly certain that the episodes of land-mammal faunal interchange, quite evident in Fig. 6–3 below, all occurred between Asia and North America and not directly with Europe. This was conditioned by changes in relative sea and crustal levels in the region of Beringia, between Alaska and north-eastern Asia, now under a shallow sea but repeatedly, episodically above sea level at various times in the past and then serving as a broad land bridge, or a corridor, for spread of land animals between the continents.

There is now a definite but limited resemblance between western European and North American land mammals with a few species and a number of genera common to the two. This results not from a direct interchange but by the disappearance of the marine barrier between Europe and Asia and the spread of various land mammals right across Europe, northern Asia, and northern North America.

Text

Related to, but distinct from, the problems of faunal interchange and migration routes that have already been discussed is the question of faunal resemblance between Eurasia and North America. Faunas may agree or differ in various ways, such as overall abundance, relative abundance of different faunal elements, ecological types represented and their relative abundance, diversity or numbers of different taxonomic groups included, etc. In the present study, however, attention will be concentrated on resemblance in taxonomic makeup, with the factors of diversity, abundance, and ecologic types minimized insofar as possible. In dealing, as in this paper, with problems of historical zoogeography, taxonomic resemblance is the most immediately significant basis of comparison, even though the other sorts of resemblances also have considerable importance.

The absolute data of taxonomic resemblance are the designated groups of mammals that the two regions have or had in common. Further study may be concerned not so much with the nature of the resemblance as with its degree. As far as determinable from the available faunas, this degree of resemblance can be measured by a proportion of groups at a given taxonomic level common to two areas to groups not common to the two. In selecting a measure of this type there are several alternatives, all of which may be derived from numerical values for N_1, the total number of known taxonomic units in the smaller fauna, N_2, this number for the larger fauna, and C, the number of units known to be common to both. The measure hitherto most frequently used has been $C/(N_1+N_2-C)$, that is, the fraction (usually given as a percentage) of common units in the total fauna of the two areas together. A careful study of the characteristics of this and of several other measures has been made, but need not be presented in full here. It has been found that the measure $C/(N_1+N_2-C)$ is quite erratic and unreliable, especially as applied to paleontological data, and that the best simple measure of taxonomic resemblance, alone, is C/N_1, the fraction of common species in the smaller of the two faunas. This measure may conveniently be used in the form $100C/N_1$, which expresses it as a percentage or index and avoids the use of fractions. Possible values of C/N_1 vary from 0 to 1 and of $100C/N_1$ from 0 to 100. This measure minimizes the effects of differences in diversity and abundance and of unequal knowledge and chances of discovery, and tends, as nearly as any available simple measure, to approximate the real population value in spite of sampling fluctuations in the data.

There are five main factors that influence taxonomic resemblance between two faunas: ecology, time, distance, connection, and source. Other things being equal in each case, the degree of taxonomic resemblance (and hence the value of $100C/N_1$) will vary (1) directly with the degree of ecological similarity between the two faunas, (2) inversely with the difference in age between the two faunas, (3) inversely with the geographic distance between the two faunas, (4) directly with the facility of communication between the two areas, (5) inversely with the extent of derivation of the two faunas from different sources.

The last three relationships have much in common and may be confused, but there is a distinction. Even in a single land area with its faunules perfectly free to mingle and with all derived, in fact, from one source, complete homogeneity of fauna is seldom attained and in general more distant faunules will differ more—this is the third of the five relationships listed. On the other hand, any restriction on faunal interchange and any derivation of faunal elements from different sources will produce taxonomic differences between separate areas—these are the fourth and fifth relationships as listed.

Interesting as the first three of these relationships are, in themselves, it is the last two that are particularly pertinent in historical zoogeographical studies like this. The first three may, from this point of view, be influences that tend to obscure the relationship that is the main object of study. In dealing with fossil faunas, particularly, it is almost never possible to eliminate any of these relationships as of possible significance in influencing the observed degree of taxonomic resemblance. Two fossil faunas can seldom be determined to be exactly similar in ecology or of exactly the same age. The influence of distance is negligible only when comparing successive faunas in the same region.

A certain rather broad degree of ecological similarity is imposed by the definition of the problem in this and most similar studies. The data are derived from nonvolant land mammals, and all these belong to a single broad ecological category, varied as they may be in regard to smaller ecological divisions. Unless these smaller divisions are of the most diametrically opposed nature, the taxonomic differences between the faunas compared are likely to be strongly marked only on lower taxonomic levels, as of subspecies or species, and to decrease or become negligible on higher levels, as of genera or families. By and large, the mountains, forests, and steppes of a single region at one time will draw their mammals from the same families and will have a number of genera in common, but probably few species and perhaps no subspecies.

A paleontological example, on a small scale but still significant and interesting, is provided by middle Paleocene faunas in central Montana. These occur in the same region and are of almost precisely the same geologic age, so that distance and time factors are negligible. Two faunules are from quarries and are closely similar in facies. Another faunule is from scattered surface finds, similar in facies among themselves but clearly distinct in facies from the quarries. The taxonomic resemblances are given in Table 6–2.

TABLE 6–2

$100C/N_1$

	Species	Genera	Families
2 quarries	73	86	100
Gidley quarry and surface localities	68	86	100

The influence of facies appears only in the figure for common species, and even here is slight. In general, such comparisons as have been made in quantitative form suggest that with appropriate samples the influence of facies among faunas of nonvolant land mammals of the same age and from the same region is less pronounced than might be expected *a priori* and is generally less than and not well distinguished from the influence of minor differences of age, geographic separation, and chances of discovery. Looked at in another way, this means that this measure of taxonomic resemblance in faunas is not a good measure of ecological resemblance and not a particularly helpful approach in ecological studies, as applied particularly to fossil faunas in which data on the lowest taxonomic levels are seldom very good. In the above examples, the figures given would not warrant the conclusion that there is any appreciable difference in facies between the quarries and the surface localities. Data on the relative abundance of the different taxonomic groups show, however, that there is a very definite distinction between these facies. Ungulates and carnivores are relatively abundant at the surface localities and rare in the quarries. Primates and insectivores are abundant in the quarries and almost absent at the surface localities. Ecological analysis, for its own sake, is most directly and hopefully approached by such study of relative abundance and not by that of taxonomic resemblance as here defined for a different purpose.

The same example may be extended to show how much more influence on this particular measure may result from geographic than from purely ecologic factors. The surface localities in Montana are rather closely similar in facies, as shown by relative abundance of the various ecological types, to the known fauna of the Torrejon formation in New Mexico, 650 to 700 miles distant. The age is nearly if not quite the same. $100C/N_1$ for species is 22, for genera 60, and for families 100. The use of families is still too broad to reveal any difference in these generally similar faunas, but the figures for genera and species are much lower than for different facies in the same region. Of course it is impossible to rule out all influence of age and facies, but the major part of this difference appears to be basically geographic, and it is suggested that even in the middle Paleocene there was a distinct faunal zoning between southern and northern faunas in what is now the Rocky Mountain region of the United States.

The measure $100C/N_1$ is quite sensitive to lateral substitution of ecological vicars, a distributional or geographic rather than facial factor which is probably the strongest influence toward lower values of this measure in the example just given. It is frequently far less sensitive to more directly ecological differences, because it deliberately avoids indicating the presence in a more varied fauna of types absent in one less varied. For instance in the comparison of middle Paleocene faunules within Montana, the presence of many primates in the quarries and their absence at the surface localities is a very striking ecological indicator, but it makes no difference at all in the value of $100C/N_1$. This emphasizes again that this measure is poor for primarily ecological studies but valuable for primarily geographic studies, the recommended use.

It must, however, be emphasized that use of this measure in this way depends on the suitability of the samples compared. With unsuitable samples, the differences summed up by $100C/N_1$ are still real, and the comparison is still valid, but it may not be the comparison intended. As with other statistical procedures, conclusions based on this measure do not flow automatically from the figures obtained but must take into consideration the characteristics of the samples and of the measure used. Also, as in all statistical methods, inferences from small samples relative to the population as a whole are less reliable than those from large samples. It is one of the advantages of this particular measure, used in the way here suggested, that it minimizes the defects of small samples and that it tends to minimize or eliminate purely facial differences, but it does so only under certain conditions.

The most important of these conditions are (1) that the larger of the two samples (known parts of faunas) compared must be relatively large, including most or a considerable part of the whole fauna, and (2) that it must be either broadly representative of a large and regional fauna, not the strictly localized fauna of a narrow ecological niche, or must be similar in facies to or must include the facies of the smaller sample compared with it. These conditions are met in the comparisons that have been discussed above, and they are met in the other comparisons to be made later in this section.

The influence of lapse of time is exemplified in an interesting way in Table 6–3, although these data do not isolate this factor from all others. $100C/N_1$ is given for genera in five successive late Paleocene to middle Eocene faunas in the Rocky Mountain region. The table clearly brings out a regular decrease in taxonomic resemblance of faunas in either direction of the time dimension—that is, progressively smaller values of $100C/N_1$ starting at any fauna and running through either older or younger faunas. The table also brings out some irregularities not in the direction but in the degree of this change. These are probably due in part to unequal lengths of time involved in the different units, but also in part to factors other than lapse of time. For instance the Gray Bull resembles the immediately following Lysite more than it does the immediately preceding

TABLE 6–3

Taxonomic resemblances, measured by $100C/N_1$ for genera, between five successive mammalian faunas in the late Paleocene to middle Eocene of the Rocky Mountain region.

	Clark Fork	Gray Bull	Lysite	Lost Cabin	Bridger
Clark Fork	100	88	53	53	6
Gray Bull	88	100	94	66	21
Lysite	53	94	100	59	22
Lost Cabin	53	66	59	100	27
Bridger	6	21	22	27	100

Clark Fork. This is probably due to difference in source (the fifth of the influences listed on a previous page). The Gray Bull fauna is not wholly derived from the Clark Fork but includes a large immigrant element of different source, while the Lysite fauna is in larger measure derived from the Gray Bull or from the same sources as the Gray Bull. There is also some geographic influence, for the faunas are not all in the same region, although not far apart. Such differences in facies as may exist apparently do not influence most of these figures much, although this is probably one element in the unusually sharp separation of the Bridger from all preceding faunas.

Returning to the main theme of this paper, comparisons will be made first of all and mainly between the whole known Eurasian and North American faunas. The only omissions in this main sequence of comparisons are animals known exclusively from what is now the Oriental Region (already zoogeographically separable in the Eocene, as has been shown) in Eurasia and from what is now the Neotropical Region in North America, although Oriental and Neotropical elements that became incorporated in the faunas of middle latitudes are included after this incorporation. Comparison of such large units is rather gross, but it is necessary, in the present state of knowledge, if general trends are to be continuously followed. Where data are available, some comparisons of more limited areas will also be made, and relationships to the broader trend noted.

For most of the Tertiary either the known Eurasian and North American faunas are fairly similar ecologically or the larger of these covers an ecological range sufficient to minimize purely ecological effects on $100C/N_1$, which is stable if the ecological type of the smaller fauna is included in the larger and does not require that both should have the same ecological range. The apparent resemblance may nevertheless be somewhat lowered by ecological factors in some cases.

Comparison is intended to be of contemporaneous faunas each, however, covering a considerable time range in itself. It is certainly not true that the spans covered are exactly the same on the two continents. This, too, may have lowered the apparent resemblance, although this complication is to considerable extent counteracted by the fact that correlation is itself in large part based on faunal resemblance, tending to make apparent resemblances maximal in faunas correlated by this method.

Geographic separation is great throughout the broad comparison of Eurasia and North America. This certainly has a strong influence on the apparent resemblance of the faunas, making all the values of $100C/N_1$ certainly much lower than they would be if only adjacent parts of the two continents could be compared throughout, which is, of course, impossible because the faunas of these regions are not known in the Tertiary. The individual comparisons remain valid as defined, however, and apparent changes will correctly reflect real trends if faunas from the same regions, however distant, are used throughout. This is not strictly true of the main sequence to be presented for Eurasia, which includes practically only European faunas until the late Eocene and does not include well-distributed faunas across middle Eurasia until the late Miocene or Pliocene. Nevertheless there is some evidence that apparent fluctuations due to different geographical locations of successive faunas are not so great as real changes in continental faunal resemblance and do not wholly obscure the latter. The known American faunas are fairly unified in geographic position, from the point of view of such broad comparisons, before the Pleistocene. The influence of inclusion of northern faunas in the

Pleistocene and Recent and some other geographic influences of the scattering of the data can to some extent be allowed for and will be discussed briefly.

The effect of facility of interchange between the two continents is one of the major factors being investigated. It clearly has had a strong influence on taxonomic faunal resemblance. However, neither in theory nor as observed is this relationship a simple one; for instance the resemblance is not a simple function of duration of a migration route but may even decrease while the migration route remains open. This involves an important series of special considerations to be briefly considered after the data have been presented.

As regards the last factor, source, there are also some peculiar elements in the present problem. Eurasia as here defined, or Holarctic Eurasia, exchanged faunal elements within itself, a more complex unit than North America, for instance in the Eocene when differentiation of a European and an Asiatic fauna seems to have been relatively great and to have resembled the relationship between two separate continents. This apparently was not true or was less true in the middle and later Tertiary, when this part of Eurasia had essentially a single continental fauna in spite of some rather pronounced local differences. But this continent also exchanged elements with southern Asia, zoologically ranking as a more or less separate continent, with Africa and with North America, during much of the Cenozoic.

North America, on the other hand, is a much smaller area and has been geographically and, as as far as the known data show, faunally more unified throughout the Cenozoic. Furthermore, after the early Eocene and until the late Pliocene, its mammalian fauna apparently had only two sources: North America, itself, and Asia. Since all the invading types during this time came through or from Asia, immigration into North America could promote only resemblance to Eurasia and not differences. Eurasia, on the other hand, drew immigrants from at least two major sources besides North America, and these could and did tend to reduce resemblance to North America. In the late Pliocene and Pleistocene North America received immigrants from another source, South America, but these never formed a very large proportion of the fauna, and they soon decreased to the present almost negligible figure.

Resemblance due to common source has a close relationship with the time element. The resemblance is instituted at the time of migration and thereafter tends to decrease with time, more passage of time, with concomitant evolution of populations on each continent, having the same effect on faunal resemblance as if the common source were becoming more remote.

Table 6–4 presents the general sequence for Eurasia and North America as wholes (with the previously mentioned exceptions of strictly Oriental and Neotropical elements), and Table 6–5 presents more scattered data involving more closely defined faunas. The data of Table 6–4 are presented graphically in Fig. 6–3.

Taxonomic resemblances in families and in genera have some similarities, but also have striking differences. Both curves begin at a relatively very high point in the early Eocene and then drop abruptly. The family curve reaches its low in the early Oligocene (dropping from 89 to 52) and then climbs irregularly to level off in the middle Miocene around a $100C/N_1$ value of about 65, which was thereafter maintained except for minor fluctuation, probably not significant except, possibly, for the slight, temporary rise in the early Pliocene. The generic curve drops still more rapidly to its low in the middle

TABLE 6–4

Cenozoic faunal resemblance between Eurasia and North America

Epoch		$100C/N_l$	
		Genera	Families
Recent		40	65
Pleistocene		28	63
Pliocene	Late	18	64
	Middle	18	68
	Early	16	71
Miocene	Late	12	63
	Middle	14	65
	Early	6	57
Oligocene	Late	7	59
	Middle	16	56
	Early	19	52
Eocene	Late	7	62
	Middle	5	62
	Early	42	89

TABLE 6–5

Taxonomic faunal resemblance of selected Central Eastern Asiatic
and North American faunas

Approximate age	Asiatic fauna	North American fauna	$100\ C/N_l$	
			Genera	Families
Recent	China and Mongolia	United States	18	52
Middle Pleistocene	Harbin	United States	58	75
Late Pliocene	"Sanmenian" ("Nihowan")	Blancan	16	58
Early Pliocene	Chinese "Pontian"	Clarendonian	11	62
Late Miocene	Tung Gur	Barstovian	17	53
Mid to Late Oligocene	Hsanda Gol	Orellan and Whitneyan	14	57
Early Oligocene	Ardyn Obo and Ulan Gochu	Chadronian	24	75
Late Eocene	Irdin Manha and Shara Murun	Uintan and Duchesnean	17	76

Eocene (5), then climbs to a secondary peak in the early Oligocene (19), drops more slowly to a late Oligocene and early Miocene low (6–7), then oscillates irregularly at a somewhat higher level (12–18) until it climbs very steeply in the Pleistocene and Recent to a high (40) not appreciably different from the early Eocene level.

Scattered data for the early Tertiary of Mongolia suggest that if Asiatic faunas for this time were better known, the general Eurasian curves would not drop so far or so rapidly but would still have about the same trends except that the family low might be later, in the middle or late Oligocene, more nearly coincident with the second generic low.

The interpretation of this part of the curves can be related to the inferred faunal

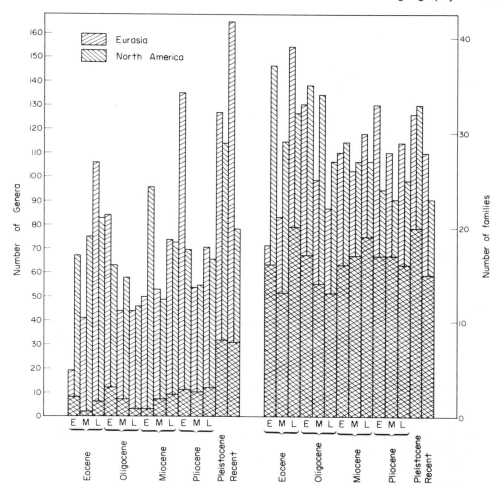

FIG. 6–3. Numbers of known genera and families of land mammals in both Eurasia and North America (double crosshatching) and in one but not the other of those land masses. Data for the Paleocene are inadequate and are not shown. The other epochs are named, and those of the Tertiary are divided into early, middle, and late parts, not scaled by time in years.

interchanges and migration routes with considerable probability and little complication. The intense early Eocene interchange established a strong basic resemblance on all taxonomic levels above the species. Interruption intermigration, which probably occurred before the end of the early Eocene and persisted until the beginning of the late Eocene, caused the faunas to evolve separately on the two continents, and this was the more marked because this was a time of rather rapid evolution among the progressive groups newly spread over the many local environments of the two continents. Resemblance declined rapidly, and this was more marked as regards genera which of course change more rapidly than do families. The late Eocene and early Oligocene interchange checked and reversed this trend as regards genera, but only slowed it down as regards families, not only because of their slower change but also because a large proportion of the families appearing separately on the two continents failed to migrate to the other. Another subsidence of the bridge, or at any rate practical cessation of interchange, in the middle and late Oligocene led to a second generic low. Probably the family low would occur here, too, on a curve corrected for geographic distribution of known faunas. That the uncorrected family curve actually rises during this time can be ascribed not only to this probable inadequacy of knowledge but also to a negative rather than positive influence in the known faunas. Increasing resemblance in their families seems to be caused rather by disappearance of families not common to the two regions than by appearance of families common to them.

The later Tertiary Eurasian faunas involved in the comparison are more widely and evenly distributed, and the general curve probably gives a fairly good representation for Holarctic Eurasia as a whole in this part of the sequence, although the minor fluctuations seen in the record cannot be considered reliable in detail. Particular Chinese faunas here tend to give figures somewhat, although not greatly, lower than for the broader comparison. The probable reason is that this region was during this time more strongly "infected" by Oriental elements, very few of which migrated to North America, than were other parts of Eurasia.

Continual interchanges during the later Tertiary brought the resemblance above the late Oligocene and early Miocene low, but it then tended to fluctuate around a moderate figure both for genera and for families. Basic differences in Eurasian and American faunas in the middle latitudes were well established, and the repeated interchanges, by way of the far North, tended to produce an equilibrium rather than an indefinitely progressive increase in resemblance. Progressive evolution in America tended to produce a drift away from the Eurasian fauna, approximately compensated by moderate migration to Eurasia. In Eurasia there was not only this sort of drift, but also incursion of some African and Oriental groups; the equilibrium nevertheless was approximately maintained by the connected factor that more groups migrated from Eurasia to North America than in the opposite direction.

This equilibrium was apparently maintained without radical change through the Pleistocene and into the Recent. There probably was some real rise in generic resemblance, but the sharp rise shown by the generic curve is in part an artifact. The reason for this is that faunas of more northern regions, where most of the common genera now occur and had probably occurred throughout later if not all Tertiary times, are here known and included in the figures for the first time. This is reflected in the very high figures for the local Harbin fauna (about mid-Pleistocene), still some distance from the actual source and reception area for migrants but much nearer to it than Eurasia as a whole and relatively although not absolutely free of Oriental elements. Figures based on all of China and Mongolia in the Recent, on the other hand, fall below those for Holarctic Eurasia in general, because they do not include so many strictly boreal elements and do include a considerable Oriental infusion.

Even from these data, with all their imperfections, it is evident that intensity of faunal interchange and taxonomic resemblance are related without being completely correlated. Faunal interchange is only one of several factors involved in such resemblance:

Positive factors, increasing resemblance:
(1) faunal interchange;
(2) extinction of autochthones in either area.

Negative factors, decreasing resemblance:

(1) progressive evolution and local differentiation in either area;
(2) extinction of migrants in one area but not both;
(3) faunal interchange of one area with a third.

All these factors are involved in the Eurasian–North American resemblance, and their separate contributions cannot easily be analyzed in the complex resultant, although the existence of such distinct contributions can be clearly seen at various points. Figs. 6–4 and 6–5 represent diagrammatically the interplay of these theoretical factors in ways related to phases of Eurasian–North American faunal history. The theoretical expectations are generalized in Figs. 6–6 and 6–7.

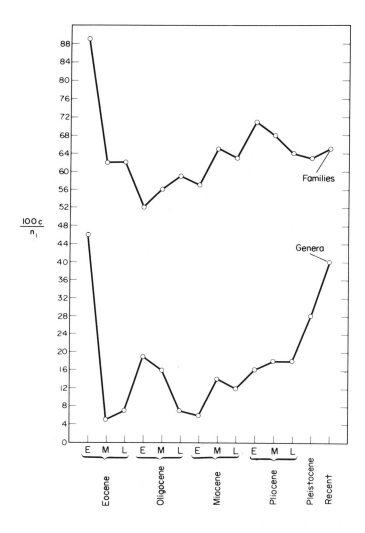

FIG. 6–4. Measures of faunal resemblance between known land mammals of Eurasia and North America through the Cenozoic except the Paleocene, measured in simfams above and simgens below.

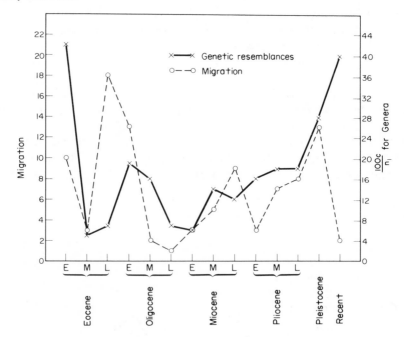

FIG. 6–5. Items of evidence for generic land mammal occurrence on both Eurasia and North America ("migration"), scaled at left, and simultaneous faunal resemblance in simgens, scaled to right.

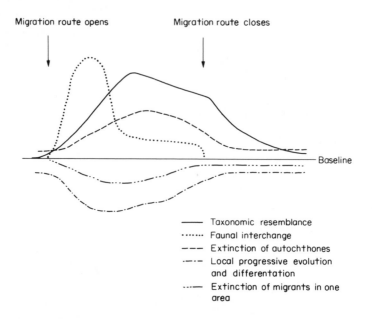

FIG. 6–6. Theoretical, generalized diagram of interplay of factors making for increase of faunal resemblance (above base line) and those making for its decrease (below base line) during one cycle of faunal interchange. This resembles the middle Eocene to late Oligocene sequence for Eurasia–North America shown in Fig. 6–5.

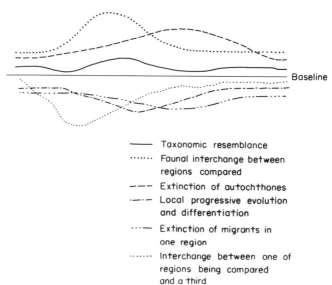

— Taxonomic resemblance

······ Faunal interchange between regions compared

— — — Extinction of autochthones

·—·— Local progressive evolution and differentiation

···—— Extinction of migrants in one region

······ Interchange between one of regions being compared and a third

FIG. 6–7. Theoretical, generalized diagram similar to Fig. 6–6 of effect of interchanges between two regions continually or continuously connected by a migration route while one also has interchanges with a third region. This is approximated by the mid-Miocene to late Pliocene in Fig. 6–5.

Intensive faunal interchange is usually followed by some extinction of autochthones, although there may be sufficient lag to be geologically perceptible. It would, then, be expected from this relationship of the two positive factors that a peak in resemblance would tend to follow a peak in interchange, but not exactly. The data for genera in the Eurasian–North American interchange show sufficient agreement with this theoretical expectation. Interchange peaks in the late Eocene, late Miocene, and early Pleistocene are followed by more or less clear resemblance peaks in the Early Oligocene, early Pliocene (present in the curve, although not well marked or clearly significant), and late Pleistocene or Recent. In each case the picture is somewhat complicated by the fact that renewed, but apparently somewhat less intense, interchange coincided with the resemblance peak.

After an important interchange, resemblance will tend to decline as local differentiation and some extinction of migrants become dominant over the positive factors, which they tend to do even if some interchange continues, and of course do more rapidly if interchange ceases. This effect is clearly seen in the Eocene and again in the Oligocene, the only times during which interchange probably virtually ceased for any considerable span. In the later Cenozoic the various factors seem to be considerably more complex, and any clear separation is then impracticable. Holarctic Eurasia received a number of migrants from Africa and the Oriental region, tending toward less faunal resemblance with North America, but this in itself seems to have stimulated some renewed exchange, tending to increase faunal resemblance, when otherwise a more stable equilibrium in interchange would have existed with slow drift downward in resemblance because of progressive evolution and differentiation in each region.

Source

(1947) Holarctic Mammalian Faunas and Continental Relationships during the Cenozoic, Bulletin of the Geological Society of America, Vol. 58, pp. 613–688. (The excerpt here presented is from pages 671–685.)

Another Reference

SIMPSON, G.G. (1943) Mammals and the Nature of Continents, *American Journal of Science*, Vol. 241, pp. 1–31. (The formula for the coefficient $100C/N_1$ was not given in this paper, but it was clearly designated in a long footnote on pp. 19–20, and figures based on it were given in a table on p. 20.)

Measurement of Faunal Resemblance

Comment

The following excerpt discusses the characteristics of some measures of faunal resemblance and some criticisms or emendations of the coefficient $100C/N_1$. Most of the first five pages of the original paper are here reprinted without essential change. The remainder of that paper discussed measures that take into account the relative abundance of individuals in the various taxa of the faunas being compared. The conclusion was that such measures are doubtfully practicable or usable for biogeographic purposes. In fact the proposed measures are not in general use, and there is little reason to reprint that discussion.

Several other indices related to faunal resemblance have been proposed, and I have commented on two of them elsewhere. One, ascribed to Preston, is:

$$x^{1/z} + y^{1/z} = 1$$

In this x is the proportion of taxa in common in one fauna and y that in the other and z is an index calculated from the equation. This produces an index not of similarity but of dissimilarity in that the higher the value of z the less the resemblance of the faunas. I have commented on this as follows:

"It is not clear to me why so complex a measure is required or why a single index of similarity . . . would not have been equally or perhaps more significant. Either the Otsuka index . . . or one suggested by me . . . could appear suitable. In these C denotes number of taxa in common, N_1 number in the smaller of two faunas, and N_2 number in the larger. The former index is $100C\sqrt{(N_1N_2)}$ and is appropriate when the faunas are of about the same size or the difference in size is not a significant factor. The latter index is $100C/N_1$ and is appropriate when sampling is considered inadequate, as is often true of paleontological samples or of zoologically poorly explored regions, or when the smaller sample may be considered an impoverished derivative or relative of the larger."

The Otsuka index is the ratio of the number of taxa in common to the geometric mean of the two faunas, expressed as a percentage. Both of the simple indices take values from 0 for no resemblance to 100 for complete resemblance. The two are equal if $N_1 = N_2$, but not otherwise.

Text

Most studies of faunal resemblances are strictly zoogeographical or correlational (in a stratigraphic sense), although other and especially ecological factors cannot be wholly excluded and are sometimes of primary importance. The usual approach is concerned with the presence or absence of taxa and with degrees of phylogenetic relationships among them. Degrees of phylogenetic affinity are practically, even though somewhat crudely, quantified by the levels of hierarchic classification. Comparisons at the specific level involve close affinity, at the generic level broader affinity, at the family level still broader, and so on. Presence and absence of taxa, for purposes of measurement of faunal resemblances, are adequately quantified by the following symbols and concepts:

E_1, number of taxa (at a specified level) in the first, smaller (or equal) of the two faunas or samples compared, but absent in the second.

E_2, number of taxa in the second, larger (or equal) fauna or sample but absent in the first.

C, number of taxa common to both.

$N_1 = E_1 + C$, total taxa in first.

$N_2 = E_2 + C$, total taxa in second.

$N_t = E_1 + E_2 + C = N_1 + N_2 - C$, total taxa in both.

The most obvious, and apparently the most intuitively acceptable measure of faunal resemblance, is simply the percentage of taxa in common among the total taxa of the two faunas or samples in questions:

$$100 \ C/N_t \tag{1}$$

which for purposes of compilation and calculation is more conveniently represented as:

$$100C/(N_1 + N_2 - C) \tag{1a}$$

If both faunas are almost or quite completely known, and if they are of at least approximately equal size, index (**1**) is usefully indicative. The maximal value, 100, shows complete indentity and the minimal, 0, complete difference at the chosen taxonomic level; and, in general, intermediate values are clearly proportionate to degrees of resemblance. If, however, either fauna is incompletely known, or if the known numbers of taxa are distinctly unequal index (**1**) is difficult to interpret and may be quite misleading.

It is generally the case with fossil faunas and sometimes with Recent faunas that the available samples, on which a measure must be based, are incomplete and do not include all the taxa of both faunas. If the two samples have the same degree of completeness relative to the populations (e.g. each has half the taxa of the respective population) and neither is biased with respect to E/C, index (**1**) from the samples may give comparable estimates of the population values. The actual ratio of sample N to population N is, however, always unknown and can rarely be closely estimated. The best one can do is to assume, in the absence of indications to the contrary, that the two populations have about the same numbers of taxa. If that assumption is true, comparisons of samples of approximately equal size still give somewhat biased estimates of index (**1**), but estimates that are reasonably comparable and are adequate if the samples are fairly large.

Those conditions are, however, rarely met. Samples are frequently of quite unequal size. It is frequently probable that the smaller sample (smaller in terms of number of taxa included) also has a smaller proportion of the taxa originally present in the corresponding population. Then the value of index (**1**) will be especially strongly biased as an estimate of the population value, and the bias will be greater the greater the discrepancy in N_1 and N_2 in the samples. The mere fact that N_2 is larger makes index (**1**) smaller. A quick and intuitively clear way to check such bias, for this or any other index, is to consider samples drawn from indentically the same population. If bias is absent, they should of course tend to give the maximal value of the index. It will at once be seen that index (1) cannot have its maximal value, 100, if the samples are of unequal size and that it may give extremely low values for samples from the same population. Obviously, comparisons based on this index are usually unreliable for unequal faunas or samples.

The following index eliminates the worst disadvantages noted for (1):

$$100C/N_1 \tag{2}$$

As an estimate of a population index from samples, index (2) minimizes effects of differences in size between N_1 and N_2. When samples are small, both (1) and (2) of course have considerable sampling error; but (2) is also preferable in this respect, and indeed the larger the discrepancy between N_1 and N_2 (the better N_2 but not N_1 is sampled), the lower the sampling error without, as with (1), introduction of increasing bias. When the sample (and population) sizes are indeed equal, (2) is still at least as good an estimate as (1), so there is no reason to turn to (1) in this (unusual) situation.

Note that (1) and (2) both have scales from 1 to 100 but that they do not tend to be equal except when (1) is 100 or either one is 0. [(1) is not, however, necessarily 100 when (2) is 100.] Otherwise (2) is always somewhat larger than (1). These, like any two different indices, measure different things, and values are to be compared only when they represent the same index.

When both faunas are completely known taxonomically, the comparison is on the true population basis rather than being based on incomplete samples. This situation is common in comparing Recent faunas of higher vertebrates, at least, but rarely arises in paleontology. When the populations are completely known, the advantages of (2) may be less obvious but may still exist. If the numbers of taxa in the two populations are equal, both indices make consistent (not identical) comparisons and there is no particular reason to prefer one over the other. If difference in sizes is in itself a meaningful element in the desired comparison, then (1) is preferable. Other things being equal, index (1) but not (2) will be smaller the greater the discrepancy between N_1 and N_2, and this is desirable if the discrepancy is important for the particular problem being investigated.

That is not, however, always or even usually the case. If, for instance, comparison is of the fauna of a smaller with that of a larger area, the zoogeographic relationships are more clearly indicated if it is possible to eliminate the tendency of larger areas to have larger faunas simply because the areas *are* larger, and then index (2) is clearly more indicative. Index (2) is even more obviously advantageous when the smaller fauna has been derived from the larger, or when a local fauna is compared with a regional fauna. The situation is comparable in stratigraphic correlation when, as is so often done, a particular faunule is compared with the whole known fauna of a given age. In these or other comparisons, it also frequently is true that the larger fauna is more varied ecologically. Then use of index (2) tends to minimize the merely ecological difference, although obviously this cannot be wholly eliminated. In effect, index (2) tends to stress the most nearly similar parts of the two faunas, which is usually an advantage in zoogeographic and especially in stratigraphic investigations. Finally, since index (2) is generally less biased for paleontological samples, it is advantageous to use (2) consistently and thus to permit more nearly valid comparisons of both fossil and Recent faunas.

Burt (1958) suggests that what is here called index (2) "should be applied in both directions to give the true picture," in other words that one should give not only

$$100C/N_1 \qquad (2)$$

but also $\qquad 100C/N_2 \qquad (3)$

This is not, in fact, the application of the same measure in two directions, but the addition of another measure with different properties. Unless the samples are equal in numbers of taxa, when (2) and (3) become identical, and except at the extreme values 100 and 0, (3) is always smaller than (2) for the same comparisons. If one thinks of these indices as truly analogous or as comparable with each other, it may appear that fauna 2 resembles fauna 1 less than fauna 1 resembles fauna 2, which is confusing. In fact (3) is a distinctly different measure that suffers to an exaggerated degree from the disadvantages of index (1), already discussed. If the discrepancy in sizes of samples is considered significant and is wanted to influence the index, then (1) is available and is more readily understood and more widely used. Otherwise, (2) is better. There seems little reason to use index (3).

Still another index that is a sort of average (but not the arithmetic mean) between (2) and (3) has been advocated especially by Pirlot (1956):

$$200C/(N_1+N_2) \qquad (4)$$

This is the percentage of common taxa not in the total for both samples as in (1), or in one sample or the other as in (2) and (3), but in the mean number of taxa for the two, as is more obvious when the formula is written:

$$100C/\left[(N_1+N_2)/2\right] \qquad (4a)$$

This is in effect a compromise between (1) and (2), for it reduces the effect of discrepancies in sample size, which is prominent in (1), but does not minimize that effect as does (2). It is hard to see a meaningful use for (4), because when discrepancy in size is pertinent to the problem (1) is better, and when it is not, (2) is better.

In contrasting (1) and (4), Burt (1958) notes that in (1) C appears in both numerator and denominator and in (4) only in the numerator. That is, however, misleading because C is really present in both numerator and denominator of all the indices (1)–(4), as is evident when they are rewritten as follows:

$$100C/(E_1+E_2+C) \qquad (1b)$$
$$100C/(E_1+C) \qquad (2a)$$
$$100C/(E_2+C) \qquad (3a)$$
$$200C/(E_a+E_2+2C) \qquad (4b)$$

If it were desirable really to eliminate C from the denominator, this could be done in some such form as:

$$100C/(E_1+E_2) \qquad (5)$$

but this is no longer a percentage with conveniently comparable values from 100 to 0, and it has no compensating advantages.

Finally, some such formula as this could be used:

$$100E_1/(E_1+C) = 100E_1/N_1 \qquad (6)$$

This is a percentage, but it is merely 100 minus index (2) and has the disadvantage of rating identity as 0 and complete difference as 100. There are, of course, comparable analogues for indices (1), (3), and (4).

An essential factor in all the indices of taxonomic resemblance is the hierarchic level of the taxa counted. It need hardly be emphasized that values of an index are comparable only if they are at the same taxonomic level and that the values will generally be higher the higher the level. Choice of too high a level for a given group of comparisons will give many or all values $= 100$, and too low a level will give many or all values $= 0$. It is almost always possible to find a level with most or all values less than 100 and more than 0, thus permitting meaningful comparisons. Determination of such a level is in itself a general indication of the degree of phylogenetic affinity involved in the comparisons.

As has been emphasized especially by Burt (1958), direct comparisons assume that the taxa really are comparable in the two faunas, that, for instance, genera are not notably more split in one than in the other. When taxa are evidently more split in one, that must be taken into account in interpreting the meanings of the indices. It may be added that index (2) has the additional advantage of tending to minimize effects of disproportionate splitting in the larger fauna.

Burt also finds it misleading that E (in the symbols here used) includes forms that have relatives (at a higher taxonomic level) in both faunas as well as those that do not. For instance in comparing the Recent North American with the Asiatic mammalian fauna, *Odocoileus* has fairly close Asiatic relatives, *Antilocapra* does not, but both are counted in E. He proposes to include *Odocoileus* (and numerous other genera with Asiatic relatives) in C. If this adjustment is clearly understood and can be made consistently, it may indeed be useful. It should, however, be noted that the same purpose can be achieved more objectively by simply including analysis at an appropriate higher taxonomic level. In all of Burt's examples his objections would have been met more or less adequately by also giving indices at the subfamily or family level.

Another possible adjustment, still closer to Burt's purpose although he does not suggest it, would be to redefine C and E:

C'_1,   Number of genera (or other appropriate specified lower taxa) in the first fauna (or sample) belonging to families (or other appropriate higher taxa) common to both.

C'_2,   Same for second fauna or sample.

E'_1,   Number of genera (or as specified) in the first in families (etc.) absent in the other.

E'_2,   Same for the second fauna or sample.

With these definitions, C', unlike C, is not a single number the same for both faunas, but indices comparable to those already discussed can readily be devised, e.g.

$$100(C'_1 + C'_2)/(2N_1 + 2N_2 - C'_1 - C'_2) \qquad (7)$$
$$100C'_1/N_1 \qquad (8)$$

Index (7) is analogous (but not directly comparable) to (1), and (8) to (2). The different form of the denominator of (7) is necessary to keep this index analogous with (1) by substituting the mean of C'_1 and C'_2 for C. A more logical but less convenient equivalent expression is:

$$100 \left\{ (C'_1 + C'_2)/2 \right\} /[N_1 + N_2 - \left\{ (C'_1 + C'_2)/2 \right\}] \qquad (7a)$$

All the taxonomic indices, (1)–(8), are based on existing classifications. They necessarily reflect such subjectivity and lack of consistency as may affect those classifications. They also necessarily neglect relationships of possible importance for zoogeography (e.g. probable direction of spread between two areas) that are not involved in or expressed by classification. Like any other statistics, they cannot automatically solve any problem. They can order and quantify the data that enter into them and thus they assist in sound comparisons. After being calculated, they must still be interpreted in the light of knowledge of exactly what it is that each of them measures and in accordance with zoological and stratigraphic principles.

Source

(1960) Notes on the Measurement of Faunal Resemblance, *American Journal of Science*, Vol. 258–A, pp. 300–311. (Most of pages 300–305 are here reprinted.)

Other References

BURT, W.H. (1958) *The History and Affinities of the Recent Land Mammals of Western North America*, American Association for the Advancement of Science, Publication No. 51, pp. 131–154.
PIRLOT P.L. (1956) Les Formes Européenes du Genre *Hipparion*, *Diputación Provincial de Barcelona, Memorias y Communicaciones del Instituto Geológico*, No. 14, pp. 1–151.

Migration Routes

Comment

Reckless biogeographers late in the 19th and early in the 20th century wound up giving land bridges a bad name. Land bridges were constructed in undisciplined minds to explain any and all occurrences of related animals in now separated regions of the earth. So many were thus imagined that when all were drawn on a map the entire earth was covered with them and no oceans were left. It was W.D. Matthew, more than anyone else, who put an end to that nonsense in the classical work cited below. There are and have been land bridges involved in dispersal or interchange of land animals, but they are and have been few and with special conditions. The Panamanian bridge now exists, but there is conclusive evidence that it did not exist during most of the Age of Mammals. The Bering bridge does not now exist, but there is conclusive evidence that it did exist off and on during much of the Age of Mammals. Almost all the other land bridges evoked by biogeographers of the old school were purely imaginary.

It was in further pursuit of those ideas following Matthew's earlier studies that I proposed a classification of migration routes into corridors, filter bridges, and sweepstakes routes in the paper here excerpted. Those concepts have now been accepted by almost all biogeographers, although inevitably there are still some dissenters. The most important change since this paper was published in 1940 has been general acceptance of the reality of continental drift. In 1940 the supposed evidence for continental drift was not merely inadequate, much of it was incorrect. In this paper I noted evidence for an early Eocene corridor between North America and Europe. Now that we have the plate tectonic evidence, it is clear that the connection was at most an intra- rather than an intercontinental corridor and at least a filter bridge. In a passage not included in this excerpt I also suggested that marsupials reached Australia by a sweepstakes route from Asia. Now it is clear that the marsupials spread, in one direction or the other, between Australia and South America by way of Antarctica. However, I still think it probable that the dispersal was by sweepstakes.

In this paper as a footnote (here incorporated into the text) I mentioned a conceivable sequence other than sweepstakes dispersal by which one-way overland dispersal could occur, but added that "there is, however, no good evidence that such a peculiar sequence of events ever actually happened." McKenna (1973) later pointed out that one-way dispersal is indeed theoretically possible by a sequence not anticipated before plate tectonic theory was developed and accepted. A part of one landmass might move away carrying part of that landmass's fauna and later come into contact with another landmass and introduce into the latter's fauna elements from the fauna of the original landmass. McKenna gave this hypothetical process the suitable name of Noah's ark dispersal.

That is certainly a possibility under current tectonic theory, but still there is no probable example of its actually occurring. McKenna specified as possible "arks": India, which is believed to have drifted from Africa to Asia; Australia and New Zealand, believed to have drifted from Antarctica to their present positions as islands; questionably the Arabian peninsula, which rifted away from Africa and to adjacent Asia; and South America, becoming connected with Central America. Although Australia and New Zealand probably had some native mammals (Australia) and birds (both Australia and New Zealand) when they left the vicinity of Antarctica, none of these are clearly known to have reached another landmass—Noah's ark is not known to have landed. South America only substituted a filter bridge for a sweepstakes route—the Noah's ark effect seems irrelevant. Arabia is still connected with both Africa and Asia—Noah's ark did not leave port. The only likely candidate for a Noah's ark effect in the late Cretaceous and Cenozoic is thus India, but as regards the mammals, at least, there is no real evidence for a Noah's ark effect. This therefore remains at present an interesting speculation with no known role in biogeography.

McKenna also proposed another possible sequence that would not involve dispersal of living animals but could confuse the evidence of dispersal. In this hypothetical sequence a fragment of a landmass, cleverly called a Viking funeral ship by McKenna, breaks off after fossils had already been deposited in it. If then it collided with another landmass, the fossil animals would falsely be assumed to have lived on the latter. Thus India, now part of Asia, might have fossils that really lived in Africa–plus–India and not in Asia. But, again, this is not known or found probably to be true.

Text

Corridors. If no barrier at all exists between two areas, it is to be expected that their faunas will be very similar, or as far as genera or larger groups are concerned practically identical. Such radical differences as exist will be mainly or wholly caused by the survival or development of local forms in some narrow environment, that is, will be facial and not geographic in a broader sense.

As an example, a comparison of the living mammals of Florida and New Mexico shows the degree of similarity attained by areas in which there is no significant geographic barrier but where the local climates and facies are almost completely different in the two areas. For various reasons not pertinent here, the mammalian fauna of Florida is relatively small, with only a quarter as many species as in New Mexico, but of the orders of mammals present in Florida, all occur in New Mexico, of the families over nine-tenths, of the genera two-thirds, and of the species nearly one-fifth. If these were fossil faunas resemblance this great (or, as is often the case, greater) would warrant the conclusions that no barrier did exist between the two. This criterion can be applied in close parallel. It was formerly sometimes supposed that when Florida first definitively appeared as dry land in the mid-Tertiary it was not yet connected with North America. Now we have from there Middle Miocene mammalian faunas with ten genera surely and six others doubtfully identified. Of these, all but one are common in contemporaneous beds in western North America. The conclusion that there was no sea or other notable barrier between Florida and these States is inescapable. Such evidence suggests not merely that a bridge existed but that none was needed; that the two areas were part of a single land mass.

Filter-Bridges. When two regions are separated by a strong barrier, they develop quite different faunas, the differences being roughly proportional to the lapse of time since the regions were connected. If now some means of passing the barrier appears, the two faunas intermingle, but usually the result is not the production of a single fauna even in the sense that Florida and New Mexico have one fauna. Several factors are concerned in the usual fact that such regions tend indefinitely after they are united still to have distinctive faunas, despite their sharing of some faunal elements. From this point of view the fact that the regions often are different environmentally exerts a profound effect, but one not of primary importance in the phenomena here considered because the effect might have been analogous even if the regions had always been united. A more important factor is that biological pressure of immigrant forms may inhibit the expansion of some groups in one region without being sufficient to cause rapid extinction, although in such cases extinction usually follows sooner or later. Equilibrium does occur but is seldom or never permanent.

Another and for the present subject a more important reason for the continued distinction of two faunas between which a barrier-crossing has been established is the character (including the position) of that crossing. Its approaches may be inaccessible for some animals, and of course they can not use a bridge that they can not reach. From the animals that do expand into a new land mass, it is sometimes possible to infer where the bridge was. Thus when North America and Asia had a great faunal interchange in the Pleistocene, no mammals then confined to southern North America reached Asia and none then confined to southern Asia reached North America. Obviously the bridge was in the north and exclusively southern animals could not reach it. It is also noteworthy that none of the mammals that had come into North America from South America then reached Asia. To reach North America they had to come through the Tropics, and none was sufficiently adaptable also to pass over a relatively cold bridge.

Here the character of the mammals themselves is a determining factor. What is a barrier for one is not for another, and conversely what is an open route for one is not for another. The Asia–North America bridge opened the barrier for elephants (mammoths) but not for gazelles. The North America–South America bridge opened the the barrier for horses but not for bison. This strongly selective action depending on the position and character of the bridge and the consequent environmental conditions of it and of its approaches is a rule with few exceptions. Another way of putting this would be to say that the true barrier in such cases was not the presence of a stretch of sea but some less obvious environmental factor, such as climate or vegetation, and that for these animals the apparent bridging of a barrier had no meaning because the true barrier remained untouched.

In the inference of intercontinental land connections from faunal relationships it is, therefore, wrong to demand that anything like a complete faunal interchange be adduced as evidence of the existence of the connection. A wide-open, nonselective connection, a corridor, is the only sort that could approach such a result, and these are rare. In the whole history of mammals there are exceedingly few cases (e.g. Lower Eocene between Europe and North America) where the evidence really warrants the inference of a wide-open corridor between two new distinct continental masses. The usual sort of connection is selective, not acting as a corridor or open door but as a sort of filter, permitting some things to pass but holding back others. From the probable mechanism of such filtering of faunas, it follows that these connections were usually of narrow environmental scope and their continental abutments limited, drawing only on one faunal zone of the continent, not on its fauna as a whole. In other words, the usual evidence for such connections does not suggest "lost continents" comprising parts of two or more as they exist today, or even broad transoceanic pathways, but relatively restricted links. The analogy of a bridge for such selective or filtering connections is fairly good, and it is to them that the term "land bridges" most properly applies.

From the point of view of paleogeography, the sort of bridge that best fits the zoological evidence in such cases of extensive but filtered faunal interchange is an isthmian link in the sense of Bailey Willis. The broad land bridges of many paleogeographers should be corridors from a faunal point of view, but isthmian links, more nearly than any other geologically postulated connections, fill the requirements of a filter-bridge, which the faunal evidence shows to be the usual type of intercontinental connection although, of course, by no means the only type.

When it is recognized that a filter-bridge does not lead to an integral transfer of continental faunas, it is a practical problem to determine what sort and degree of resemblance does indicate such a bridge. There have been students who did not hesitate to build extensive individual bridges in all directions to account for peculiarities of distribution in single forms of life. Thus, to mention only a few of his many connections, Joleaud has an individual late Oligocene route from Haiti to west-central Africa for insectivores, one diagonally across this from Brazil to northwestern Africa in the Late Eocene for certain rodents, one in the Early Miocene straight across the Atlantic from the United States to Spain for a genus of horses, *Anchitherium*, one at the same time parallel to but south of this from northern Africa to Florida for the mastodonts, and so on. Similarly, von Ihering built a special bridge across the Pacific from South America to Asia for raccoons and bears, and examples could be multiplied. Aside from geological considerations, which in themselves are almost enough to exclude these particular bridges at these places and times, and aside from what are now known to be errors in the factual data adduced for them, such individual, self-service bridges are supposed to have acted in a way in which no surely established bridge is known to have acted, and I can not believe in their reality.

One good criterion of the reality of a bridge is that it should have acted in both directions. Provided that both areas had land faunas, there seems to be no proved case in which a bridge has conducted animals only from one to the other and not in both directions. This is true even when one fauna was decidedly dominant and tended as a general rule to suppress the other or to inhibit its expansion. For instance, the South American ground sloths were doomed to extinction when they came in contact with the North American fauna, but first they penetrated far into North America. The armadillos, also archaic animals such as might be expected to contract in distribution, have gained an even more enduring foothold in North America and are now (for at least the second time) expanding there. One of the best arguments against the disputed derivation of South American marsupials by land bridge from Australia (direct or via Antarctica) is that the evidence favors migration only from Australia to South America, with none in the reverse direction even though the South American mammals must have been at least as capable of expansion as the Australian. It is conceivable that a bridge might function in one direction by a sort of lock or storm-door action, an otherwise uninhabited region receiving a fauna first from one source, losing that connection, and only then being united with a second continent, so that animals would be transported from the first to the second but not in the other direction. There is, however, no good evidence that such a peculiar sequence of events ever actually happened and it should hardly be postulated except in the absence of any acceptable alternative hypothesis.

The second and perhaps the best criterion of the reality of a land bridge is that even though it rarely transports whole faunas, it does tend to transport integrated faunules. It does not transport all the genera of a continent, but neither does it transport one genus all by itself. For instance, it is improbable that only herbivores or only carnivores would cross such a bridge (although they need not both cross in the same direction). Where herbivores go, carnivores can and will accompany them, and carnivores can not go where there are no herbivores. The postulation of land bridges on the basis of one or a few mammals is thus very uncertain. Unless there is reasonable possibility that there companions have not been discovered, a theoretical bridge based on such evidence is probably unreal.

Sweepstakes Routes. There are, however, instances of migrations of single groups of mammals or of unbalanced faunas that did occur but that do not meet these criteria for filter-bridge connections and, of course, still less those for corridors. Many insular faunas are of this type, as a whole. Madagascar and the West Indies are classic examples. As carnivores, Madagascar has only peculiar viverrids, relatives of the civets, although nearby Africa is abundantly provided with cats large and small and various other carnivores. Madagascar's insectivores and rodents are also peculiar and each group is related to only one of many African types. Madagascar has many primitive primates and lemurs, but no apes or monkeys. These are all ancient forms and constitute a very unbalanced fauna that must have entered (whether together or separately) by the middle Tertiary at latest. The only ungulates are a pigmy hippopotamus (now extinct) and a bush-pig, both of which must have reached Madagascar much later than its other mammals and which are, again, an example of migration that can not possibly be explained by an ordinary filter-bridge. In the West Indies the Pleistocene land mammals included only peculiar rodents, insectivores, and ground sloths, without any of the ungulates, carnivores, and other groups abundant on all adjacent continental areas. This fauna, too, is inexplicable as a result of normal filtering on a land bridge such as is here envisioned. I am aware that some excellent authorities do maintain that these faunas arrived over bridges, but I can not feel that they have clearly seen or considered the conditions that could give such a result.

There are also instances of the appearance of isolated immigrants on continental masses. A curious and relatively neglected example, among many that might be cited, is that of the sudden appearance in South America of small relatives of the North American raccoon. These procyonids appear as fossils in the Late Miocene or Early Pliocene of Argentina definitely before any of the other carnivores or any of the abundant North American ungulates reached there. Since in this case a filter-bridge certainly existed at a later time, it is usual to assume that the procyonids came on this bridge and that their appearance dates the formation of the bridge as a practicable migration route or true and complete filter-bridge. If, however, we consider only the time when the procyonids did appear, disregarding our knowledge of what was destined to happen later, such a conclusion is not warranted. If my previous remarks as to filter-bridges are true, or are acceptable as a theory of general tendencies, then it is wrong to conclude that a bridge can account for the appearance of this one group of small carnivores and no other animals of similar geographic origin at that time, unless the bridge was then so nearly impassable as not to warrant the name in its usual accepted sense.

The late W.D. Matthew, who was probably the most distinguished and best informed student of problems like this, concluded that insular and highly unbalanced faunas were probably to be accounted for by sporadic transportation of land animals on natural rafts, without the existence of a dry-land route (Matthew, 1918, 1939). This opinion has been severely criticized in some quarters. It has been claimed or felt, even by some adherents of Matthew's general thesis of "Climate and Evolution," that this sort of adventitious migration is dragged in when necessary to explain away any facts that contradict the main thesis.

It has not been sufficiently emphasized even by Matthew that the role of such a theory may be positive and primary, not merely negative and supplementary. Adventitious migration has indeed been used and sometimes abused simply to get inconvenient facts out of the way of a favored hypothesis, but there are instances in which adventitious migration is itself the most probable hypothesis and the most economical theory. In the cases of the faunas of Madagascar and the West Indies, for instance, I strongly favor this explanation, and I do so not at all in order to explain away data for a land bridge where I do not want to believe in one—as Matthew has, quite incorrectly, been accused of doing. It is to be favored because it does explain, simply and completely, facts that the land-bridge theory does not explain.

This sort of migration can be extended to include cases other than those of transportation by natural rafts, although doubtless these provide the most common instances. Any barrier, whether of water, climate, biota, or other, may or will be involved in such migration if its crossing at any one time is highly improbable but is not impossible. The action is not merely like that of a relatively less permeable filter but is different in kind as well as in intensity. A filter-bridge permits some animals to pass and holds others back, but in general those that can cross it do cross it and do so fairly soon after the bridge becomes available to them. It is relatively deterministic as to the fact of crossing, as to the animals that do or do not cross, and as to the time of crossing. An adventitious route, which I call "a sweepstakes route" to emphasize this characteristic, is indeterministic. Its use depends purely on chance and is therefore unpredictable and, except in a broad way, can not be clearly correlated with other events in time and space, as filter-migration can.

That such views have not received much attention and that they are uncongenial to many zoogeographers is perhaps a reflection of the mechanistic scientific philosophy dominant in the Victorian age, from which zoogeography has not fully emerged. Land-bridge migration seems more mechanistic because it is often more simply predictable. In fact, of course, it too depends on chance, but here on the chances of a probable event, whereas sweepstakes migration depends on the chances of an improbable event. The viewpoint involved is, I believe, new, and it merits detailed consideration, but this can not be given here. Among other points, the physical nature of such sweepstakes routes needs study. It is not to be supposed that they are invariably island stepping-stones or that natural rafts are the sole means of transport involved.

Source

(1940) Mammals and Land Bridges, *Journal of the Washington Academy of Sciences*, Vol. 30, pp. 137–163.

Other References

MATTHEW, W.D. (1915) Climate and Evolution, *Annals of the New York Academy of Sciences*, Vol. 24, pp. 171–318.
MATTHEW, W.D. (1939) *Climate and Evolution*, Second edition, revised and enlarged, Special Publication of the New York Academy of Sciences, Vol. 1, pp. i–xii, 1–223. (A posthumous reprint of Matthew [1915] with added matter by him and others.)
McKENNA, M.C. (1973) Sweepstakes, Filters, Corridors, Noah's Arks, and Beached Viking Funeral Ships in Paleogeography, in *Implications of Continental Drift to the Earth Sciences*, edited by D.H. TARLING and S.K. RUNCORN, Academic Press, London and New York, Vol. 1, pp. 295–308.

An Example of Sweepstakes Dispersal

Comment

Darwin's early account of his visit to the Galápagos Islands in 1835 is not clear on this point, but it is evident that thereafter he was often thinking about the spread of non-marine plants and animals to such islands. In 1855, several years before publishing anything about evolution, he did publish a short note on experimental evidence for possible transport of viable seeds over long stretches of ocean. In the first edition of *The Origin of Species* he treated the "Inhabitants of Oceanic Islands" at considerable length (18 pages). His principal points were that in general the nonaquatic biotas, both floras and faunas, of oceanic islands had reached them by what I now call "sweepstakes dispersal," several means of which were mentioned by Darwin, and that they are usually related to the biotas "of the nearest mainland, without being actually the same species" (Darwin, 1859).

Darwin noted that oversea transport would seem particularly difficult for land snails, which are abundant in Hawaii and other oceanic islands, but he indicated both observational and experimental evidence that such dispersal was quite possible for the land snails. Nevertheless almost a century later a leading authority on land snails, H.E. Crampton, was still insisting that they could have reached Hawaii only overland. Most recent biogeographers, however, assume that most or all really remote and isolated islands, among notable examples the Hawaiian and Galápagos Islands, received the ancestors of their biotas by sweepstakes dispersal.

There are still special cases that continue to cause some uncertainty and disagreement. Sweepstakes dispersal of nonaquatic biotas is less probable than dispersal by corridors or filters, and it may be difficult to grasp that the improbability of dispersal may itself be the most probable explanation for the actual character of some island biotas. An outstanding and long-disputed example is the mammalian fauna of the Greater Antilles, this aspect of which is discussed in the following excerpt from a longer paper in which the mammalian fauna was further treated in detail as to its composition, geographic origins, times of dispersal, and spread among the islands. In the following text I have here omitted those details but have included from the end of the paper a discussion of the zoogeographic status of the Greater Antilles as a whole.

A somewhat similar but more complicated zoographic situation is provided by the islands of the Malay Archipelago that lie between the borders of the Asiatic and the Australia–New Guinea continental shelves. I have recently reviewed previous studies and the available evidence and decided for the East Indies not on the continental shelf, as for the West Indies, that they cannot reasonably be placed in either of the two adjacent major faunal regions (Oriental and Australian) and are too heterogeneous to be defined as a region in themselves. They should be left out of the classical system of regions and studied on their own. (That study is cited on p.229, and a map prepared for it is reproduced on p.251.) That treatment of the West Indies and of the East Indies has not been willingly accepted by some biogeographers, but neither have the evidence and reasoning been restudied by them.

Text

Earlier zoogeographers mostly assumed that the presence of a rather rich terrestrial biota on the Greater Antilles required the former existence of one or more land bridges from continental areas to the islands. We owe to the genius of Matthew (1915, 1939) the first really serious and reasoned suggestion that the Antillean fauna not only is consistent with introduction by over-water routes but also is *better* explained by that theory. The over-water theory was heavily attacked by Barbour (1916), by Scharff, and by others. It was strongly sustained by still others, notably by Myers (1938) and by Darlington (1938), who reviewed the whole subject with great care and demonstrated that all the objections to over water introduction of the Greater Antillean fauna are clearly invalid. Since 1938 a majority of students who have made any considerable personal study of the subject have agreed with the essentials of Matthew's views.

When I first drafted the present paper it seemed useless to reopen a question that had been so thoroughly discussed and had, I felt, been settled. Yet it appears that many students still consider this the principal general reason for interest in the Antillean fauna. One of the most recent textbooks of zoogeography (de Beaufort, 1951) not only considers the question still open but also strongly inclines toward the land-bridge theory. It may therefore be useful, if not necessary, to summarize the arguments again.

The main arguments against over-water dispersal and in favor of land-bridge connections are as follows. I have added in brackets a brief statement of the counter-arguments.

(1) Over-water dispersal of so many, or of these particular, terrestrial organisms is extremely probable. [It can be demonstrated that the nature of the fauna is *best* explained by the postulate that colonization was, indeed, highly improbable.]

(2) Such dispersal for some Antillean animals is not merely improbable but definitely impossible. Barbour cites, for example, burrowing amphibians, onychophores, and cyprinodont fishes. ["Impossible" expresses only an opinion that can easily be shown to be fallible in these and other supposed examples. Darlington and Myers have conclusively demonstrated that over-water dispersal is a real possibility for all the groups of organisms that do, in fact, occur in the Greater Antilles.]

(3) "The islands of the Antillean chain have too evenly distributed and homogeneous a fauna for it all to have been fortuitously derived." [Actually the faunas of the various islands are remarkably heterogeneous. They are striking precisely for their comparative lack of homogeneity. That the distribution is orderly is quite a different point and is, as Darlington (1938) stressed, at least as consistent with over-water dispersal. "Fortuitously" is a misleading word, because low-probability dispersal still is not truly fortuitous in the usual sense of that word.]

(4) There are too many different elements in the fauna for all to have come by "'flotsam or jetsam' dispersal" (Barbour, 1916). The amphisbaenans, for instance, are said to require eight "practically inconceivable voyages" (*ibid.*). [Quite the contrary: an outstanding peculiarity of the Greater Antillean fauna is that it includes comparatively few groups, many fewer than continental islands or areas on the continents of comparable size and ecological diversity. The necessary number of introductions has also been grossly exaggerated; the amphisbaenans, for instance, require at most three introductions, not eight.]

(5) If mammals are rafted at all, they should be far more common on the islands than they are. [That simply reverses the logic of the situation, besides flatly contradicting the previous argument, although Barbour adduced both. The actual abundance of mammals reflects a certain low degree of probable dispersal. They should, indeed, be more common (diverse) if they came on a land bridge.]

(6) The supposedly few animals that might really be transported by rafts include some that are conspicuously absent in the West Indies, for example, the "almost semiaquatic" *Basiliscus* (Barbour, 1916). [*Basiliscus* is a poor choice, because rafting is really highly improbable for it (Hecht, personal communication). On the same basis it is, indeed, strange that didelphids or procyonids did not reach the Greater Antilles. But if a land bridge existed, these same failures are far harder to understand than on the theory that colonization was highly improbable in any case and happened to fail in these instances. *Procyon* did reach the Bahamas and the Lesser Antilles, almost certainly by relatively recent rafting from, respectively, North and South America.]

(7) The very large percentage of endemic species on each island would not have arisen if "a constant exchange of individuals [by over-water dispersal] from one island to another were taking place." [Again the facts really better fit the theory of over-water dispersal, which does not involve "a constant exchange" but comparatively rare incidents. A single introduction of one or a few animals could give rise to an isolated colony soon as large as the ecology permitted. Subsequent occasional arrival of related strays would have virtually no influence on the evolutionary divergence of the established population.]

(8) "Rafting from island to island could certainly not have occurred" (Barbour, 1916), yet the smaller islands have merely an attenuated representation of groups that also occur on the larger islands. Why should just those groups and no others have succeeded in over-water transport from the mainland repeatedly to larger and smaller islands? [It is agreed that in most cases, if not all, the few mammals of the small islands were derived from the nearest of the four main islands. This would be likely whatever the means of dispersal and has little or nothing to do with how the animals reached the main islands. Over-water dispersal from island to island is, furthermore, by no means ruled out.]

(9) The islands may well have had a balanced fauna of continental type that has simply been impoverished by the vicissitudes of island life. The ungulates, for instance, may have been exterminated by repeated flooding of the lowlands. [The fauna as a whole is rather rich. It is poor only in comparison with areas that did have land connections. It does not seem to be impoverished but seems simply to have evolved from comparatively few ancestral species. It is, as Darlington says, an accumulation, not a residue. The mammals, to be sure, are now impoverished but that occurred quite recently, in large part since man reached the islands. It seems, moreover, virtually impossible to imagine what vicissitudes of island life could have eliminated all the carnivores, for instance, while suitable prey for them continued to be abundant.]

So much for the principal arguments against over-water dispersal and the reasons why those arguments do not weigh heavily at least. Here are some of the most impelling positive arguments in favor of over-sea dispersal:

(1) The fauna represents a remarkably small number of major groups or ancestral stocks in comparison with faunas on adjacent mainlands. The absence of so many of the most abundant and ecologically suitable mainland groups is inexplicable if there was a land bridge.

(2) The comparative poverty of the fauna cannot be explained by insular conditions. Trinidad, for instance, has a much richer fauna, although it is smaller and no more varied than any one of the main islands of the Greater Antilles (Darlington, 1938). The most reasonable explanation is that Trinidad was and the Greater Antilles were not accessible by a land route.

(3) The stocks that are present all belong to groups for which over-water transport (by a variety of different means) is a distinct possibility. Birds, bats, and insects are especially varied and are demonstrably apt at over-seas colonization. Similar stocks of these groups occur on innumerable islands certainly never reached by land bridges. The probabilities vary greatly from group to group, but average probability of over-water transport is clearly much greater for the island stocks than for the whole faunas of adjacent continents. That dispersal was indeed over water is the only reasonable explanation for this difference.

(4) Groups for which any form of over-water transport is really extremely unlikely,

such as the ungulates among mammals, are indeed absent. There are no really large animals. Contrary to statements widespread in the early literature, there are no strictly fresh-water fish.

(5) The fauna is ecologically unbalanced. The most striking example is the fact that the mammals include no carnivores. There are other niches abundantly filled on the adjacent continents and potentially present on the islands but empty in the latter. It is virtually inconceivable that a land bridge could have filtered out all carnivores and all members of other niches empty on the islands. Over-sea transport, with its low overall probability and its random selection of groups of equal probability, is the only reasonable explanation.

(6) It is improbable that the fauna came from one geographic source. The point is debatable and some authorities think most, if not all, came from Central America. If that is correct, the land-bridge theory must visualize at least three bridges, each of which was crossed by just a few stocks. That multiplies the previously mentioned difficulties to such a point as to border on absurdity.

(7) There is evidence that the ancestral stocks reached the islands one or a very few at a time, scattered through a geologically long span. The land mammals, for instance, seem to have evolved from a few (probably nine, possibly one or two more) separate original colonizations during 30 million years or probably even longer. That is characteristic of waif or sweepstakes dispersal. It is incomprehensible in terms of land bridges.

(8) There is an extraordinary degree of endemism in each of the various islands, still more in the islands as a whole in comparison with the mainland. Long separation of the large islands from one another and of all from the mainland is indicated. Nevertheless the degrees of endemism vary greatly in different groups. There are endemic families, but there are also nonendemic genera. It is necessary to conclude that some groups were isolated on particular islands while other groups were not isolated on those same islands (or reached those islands while the other groups were isolated there). Again this is just what would be expected of waif dispersal and is virtually inconceivable with any system of land bridges.

It may be added that ocean currents and hurricane tracks are and probably have long been favorable to the rare introduction of waifs to the islands. (On storm tracks, see Darlington, 1938.) It may again be emphasized that rare colonization over a long period of time best fits the facts.

The direct geological evidence is mostly neutral: it neither suggests nor strongly negates the possible presence of land bridges. Geologists who have indicated such bridges on paleogeographic maps have done so entirely on what they believed to be the zoological evidence. The strictly geological evidence does not suggest the presence of land bridges.

On the basis of present evidence, the over-water dispersal theory is so much the more probable that the land-bridge theory is not worthy of further serious consideration except in the unlikely event of the discovery of wholly new evidence favoring it. This does not exclude the possibility of some past bridges between islands within the group. However, the land mammals, at least, seem on the whole to give evidence against even that possibility.

Status as a whole. Virtually all zoogeographers have followed Sclater in including the West Indies with Central (= Tropical North) America and South America in Neogaea, or the Neotropical region. Only recently has any important dissent appeared. Schmidt would associate the Greater Antilles with North America in the Holarctic. His reasons are (1) that the Greater Antillean fauna is closest to that of Central America, and (2) that Central America belongs historically with North and not South America. It is demonstrated herein that (1) is not true of the land mammals. Regarding (2), the statement is probably correct for most of the Tertiary, at least, but it is not pertinent to the present topic for two reasons. First, as regards the West Indies it would be pertinent only if their fauna were also historically connected with North America. That is certainly not true of the bulk of the land mammals. All of those except the two genera of insectivores and *Oryzomys* [rice rats] (only 12% of the known genera) are obviously of South American origin, whether they reached the islands by way of Central America or more directly. Second, the only possible objective way to define zoogeographic regions is by their faunal characteristics at a particular time. That Central America once upon a time had a purely North American fauna does not alter the fact that its fauna now resembles that of South America more than it does that of temperate North America. It was in a Neartic region, but it is now in the Neotropical region.

Table 6–6 shows the regional resemblances of the Greater Antillean land mammalian faunas. The resemblance in terms of genera are hardly significant because of the extreme generic endemicity of the Greater Antillean fauna. The comparison by families shows much less resemblance to the Nearctic region (temperate North America) than to the Neotropical region (South America plus Central America or more exactly tropical North America). As between Nearctic and Neotropical regions, the classic zoogeographers were obviously right in putting the West Indies in the Neotropical, as far as the land mammals are concerned.

TABLE 6–6

*Faunal Resemblances of the Greater Antilles
and the Mainland*

	South America	Tropical North America	Temperate North America
100C/N$_1$ for families			
Jamaica	80	60	20
Cuba	60	40	20
Hispaniola	50	33	17
Puerto Rico	50	50	25
Whole group	67	44	22
100C/N$_1$ for genera			
Jamaica	14	14	14
Cuba	0	0	0
Hispaniola	0	0	0
Puerto Rico	12	12	0
Whole group	8	8	4

Within the Neotropical region, the Greater Antillean fauna is distinctly more like that of South America than that of Central America, contrary to a widespread impression. This is even true of Jamaica, which has the greatest resemblance to Central America. In fact, however, the resemblance is far from close in any case. The maximal resemblance of the whole Greater Antillean fauna to any mainland faunas is less than the resemblance of some local faunas almost universally placed in different regions, for instance those of Colorado and of Neotropical Mexico. The latter comparison has the index $100C/N_1$ for families 67 and for genera 43. Even between Colorado and the Guianas the generic index is 16, twice as large as between the Greater Antilles and South or Central America. Apart from flat measures of resemblance, the whole makeup of the West Indian fauna is highly peculiar and extraordinarily unlike that of any mainland region.

The fauna of the Greater Antilles is indeed so distinctive that on this basis that island group fully merits designation as a separate major faunal region all by itself. But it is also an extremely heterogeneous region. The same measures applied to faunal resemblance among the four major islands give values that are also remarkably low, quite within ranges of some resemblances between established separate major faunal regions. Thus each island merits, if judged on this basis, designation as a separate major faunal region. This tends towards a *reductio ad absurdum*.

In fact the whole concept of faunal regions, although an excellent descriptive device when applied to continental areas that do in fact share large regional faunas, simply is not applicable to many smaller areas such as those of the Greater Antilles. They do not now have and never have had a balanced regional fauna such as those on which the classical faunal regions are based.

Some islands clearly are populated by more or less attenuated marginal parts of a continental fauna. They can then usefully be referred to the corresponding faunal region. This is true of Trinidad and even of the Lesser Antilles, the few land mammals of which are almost all truly Neotropical. The British Isles, Japan, Sumatra, and New Guinea are examples in other faunal regions. Other islands have not merely attenuated faunas but faunas that are unbalanced or ecologically incomplete, that include few major groups (e.g. orders), that are composed of elements of apparently quite distinct ages, and that have an extraordinary degrees of endemism at levels from genera upward. The Greater Antilles and Madagascar are the classic examples for land mammalian faunas. Beyond the reach of mammals, the Hawaiian Islands and the Galápagos are other classic examples for their birds and some other faunal elements.

A few doubtful cases exist, but most islands can be put unequivocally in one of the other of those two categories. This fact has of course been noticed long since, sometimes as a distinction between continental and oceanic islands, although the zoogeographic categories do not quite coincide with usual geological definitions. The islands with attenuated continental faunas are, for the most part, those that were connected with a continent during the Pleistocene. That is not, however, invariably true. The Lesser Antilles (excluding Trinidad) were almost certainly not connected with South America in the Pleistocene but do have an attenuated continental Neotropical fauna. New Britain and the Solomon Islands seem to be another example of an attenuated continental (here Australian) fauna without a Pleistocene land connection. The crucial factors in the quite different cases such as the case of the West Indies seem to be the long existence of a major water barrier which has nevertheless been crossed by separate faunal elements scattered through a geologically extensive span of time. It is, of course, further necessary that the islands in question must have been continuously above water for the pertinent span of time.

As to the zoogeographic classification of such isolated island units with abnormal faunas, there seem to me to be only two fully logical alternatives. We can recognize that each such island, or in some instances group of islands, is in fact as distinctive as one of the classical continental zoogeographic regions. It has been pointed out that this alternative would almost demand recognition of Jamaican, Cuban, Hispaniolan, and Puerto Rican regions, each coordinate with the Nearctic and Neotropical regions. The other alternative is to confine the designation of regions to the continents and the islands that really share their faunas and to exclude the other islands from the regional system. This may not seem tidy to those who feel that every spot of land must be included in some defined zoogeographic region, but it is logical, and it not only fits but also helps to explain the facts. Preferring this alternative, I would consider the Greater Antilles a special case, to be considered as such on its own merits. These islands belong neither to the Neotropical nor to the Nearctic region. That their land mammals are mainly derived from the Neotropical region is an interesting and important point, but it has nothing to do with the descriptive designation of the recent zoogeographic situation. The Recent Neotropical fauna is mostly derived from the Nearctic, but it would not be useful to conclude that the Neotropical region should therefore be included in the Nearctic.

Finally it may be observed that the Greater Antilles are a zoogeographic dead end. They have not served as a dispersal route between any other regions, nor have they been a source for land mammals in any other faunas. The latter point may at first sight seem possibly contradictory of some of the theses of this essay. If there was over-water dispersal from South and Central America to the Greater Antilles, why was there none in the opposite direction or to southeastern United States? Currents and storm tracks favor one-way transport from South or Central America to the Greater Antilles. They by no means preclude transport to southern United States, but here there is the additional factor that until the end of the Pleistocene the over-water distance was much greater than at present. In all three cases there is the crucial fact that North, Central, and South America have been continuously occupied by large, balanced, and ecologically essentially complete or closed faunas. The chances that waifs from the islands would survive in such communities are almost negligible. On the other hand most if not all waifs (at least those of just such stocks as did colonize the islands) would find in the Greater Antilles what was essentially an ecological vacuum for them. Once the hazardous trip was over, survival and expansion would be much more likely than not. That mainland–West Indian colonization was mainly if not wholly one-way is thus readily explicable on the theory of waif transport. Incidentally, this would be far less likely if there was a land bridge, and it is another bit of evidence against that theory.

Source

(1956) Zoogeography of West Indian Land Mammals, *American Museum Novitates*, No. 1759, pp. 1–28. (The excerpts here given are from pages 1–11 and 21–26.)

Other references

BARBOUR, T. (1916) Some Remarks upon Matthew's "Climate and Evolution," with Supplementary Note by W.D. Matthew, *Annals of the New York Academy of Sciences*, Vol. 27, pp. 1–15. (Reprinted in MATTHEW, 1939.)

BEAUFORT, L.F. DE (1951) *Zoogeography of the Land and Inland Waters*, Sidgwick and Jackson, London.

DARLINGTON, P.J., Jr. (1938) The Origin of the Fauna of the Greater Antilles, with Discussion of Dispersal of Animals over Water and through the Air, *Quarterly Review of Biology*, Vol. 13, pp. 274–300.

DARWIN, C. (1859) *On the Origin of Species by Means of Natural Selection, or the Preservation of Favoured Races in the Struggle for Life*, John Murray, London. (The discussion of oceanic islands is on pp. 388–406; a facsimile of this first edition was published by Harvard University Press in 1964.)

MATTHEW, W.D. (1915, 1938) (See p.221 of the present book.)

SIMPSON, G.G. (1977) Too Many Lines: The Limits of the Oriental and Australian Zoogeographic Regions, *Proceedings of the American Philisophical Society*, Vol. 121, pp. 107–120.

Probability and Time

Comment

In 1949 a conference on possible former land connections across the South Atlantic was held at the American Museum of Natural History. Its extensive proceedings were published in 1952 and the conclusions were, on the whole, quite inconclusive. Much of the discussion involved paleo- and neo-zoological and paleo- and neo-botanical evidence. Diverse and controversial opinions were expressed about the probability of animal and plant dispersal and the possibility of transoceanic dispersal as against that of a land connection. The discussion of the relationship between such probabilities and the lapse of time is the topic of the part here excerpted from my contribution to the symposium.

In order to focus on a definite item used as biogeographic evidence I related estimates of probability to the phytogeography of *Rhipsalis*. This is a large genus of cacti with at least 57 species considered valid in a standard work on that group. The true cacti are otherwise restricted to North and South America as native plants, but *Rhipsalis* had been reported as native also in Africa and in Ceylon (Sri Lanka). One botanist (Roland–Gosselin) recognized eight species as growing wild in the Old World, but considered them all identical with South American species and believed that all of those or their seeds had been flown across the South Atlantic by birds. Another botanist (W.H. Camp), however, rejected that as a "glib explanation" and insisted that the distribution of *Rhipsalis* proved that a land connection must have existed not long ago, geologically speaking, between South America and Africa. A zoologist (H.E. Anthony) who was also an ardent horticulturalist insisted that if bird transport were possible at all it should have resulted in even more colonizations of *Rhipsalis* than were known to have occurred, but that in any case the chances were that African *Rhipsalis* which is often grown as a house plant, represented merely escapes from cultivation.

Not being either a botanist or a horticulturalist, I refrained from opting for bird, land, human, or other transport for African (and Sri Lankan) *Rhipsalis* but indicated that there had been a lot of unnecessarily loose guessing about probability in such discussions. It did not strike me or anyone else at the time but does occur to me now that it would have been appropriate to ask why several (eight according to Roland–Gosselin) species of *Rhipsalis* and none of the hundreds of other species of New World cacti got to the old World and now seem to be native there.

I have here excerpted only the method and not the example. In this day of programmable pocket calculators and easy access to computers, my suggestion that calculating q^t could most easily be done with logarithms is decidedly dated, but the method is still valid and anyone using a program for a calculator or computer should still know exactly what that program means in the mathematical terms used here.

Text

For any event, there is a definite probability of occurrence. In flipping a coin, the chance of getting heads is one in two, or the probability is one-half. In simple mathematical conventionalization, $p = 0.5$. This is an exact, true statement of probability (if the coin is theoretically perfect). It is true, however, only if the coin is flipped once and no more, just as it may be true that spread of organisms across a barrier is improbable only if we think of a single individual or of one population over a short period of time such as a year of a century. If the coin is flipped 10 times in succession, or (what amounts to the same thing as far as probability goes) 10 coins are flipped once, then the probability of getting heads at least once is far greater than one-half, or 0.5. In fact, it is greater than 0.99, and a gambler would be justified in taking this as a sure thing. Statements of probability in coin-flipping mean nothing unless they take into account the number of trials or opportunities for occurence. Similarly, statements of probability of dispersal are meaningless unless they take some account of analogous effects tending to change the value of total probability.

Probability of dispersal is affected in this way by the size of the population subject to dispersal. If the chance that any given individual will cross a geographic barrier is, say, one in a million, $p = 0.000001$, then in a population of a million individuals the chance that at least one will cross the barrier is almost two in three, $p = 0.63$. An event so extremely improbable for a single individual that most of us would think of it as impossible becomes more likely than not for the population as a whole. The passage of time has the same effect: it multiplies the opportunities for occurrence, whether one considers this as giving the same individuals (or their successors) more opportunities for dispersal or as increasing the number of successive individuals that have such opportunities. If the probability that some member of a population will cross a barrier is 0.000001 in any one year, in a large population this means that the probability for any one designated individual is almost infinitesimally small, so much so that it would seem absolutely impossible to even the best qualified observer in the field. Yet during the course of a million years the event would be probable, $p = 0.63$, again. In the course of 10 million years the event would become so extremely probable as to be, for most practical purposes, certain, $p = 0.99995$. In other words, there would be only about one chance in 20,000 that it would not occur. One million years is not a particularly long time, as time goes in dealing with many problems of historical biogeography. This example, in itself, should give pause to those who speak of the "improbability" of transoceanic dispersal on the basis of their observations of a few individuals for a few years.

The calculation of such probabilities is fairly simple and is explained in innumerable books touching on statistics or probability, but it is not familiar to all biologists, so may be briefly explained here. The symbol for probability of occurrence is p, and that for probability of nonoccurrence is q. Values of p lie between 1 , certainity of occurrence, and 0, impossibility of occurrence. The value of q is always $(1-p)$. The probability in multiple trials or opportunities for occurrence is obtained by expanding the binomial expression $(p+q)^t$, where t stands for the number of trials or opportunities. For instance, for two trials this expression becomes

$$(p+q)^2 = p^2 + 2pq + q^2$$

The first quantity, p^2, is the probability that the event will happen twice in two trials. The second quantity, $2pq$, is the probability that it will happen once, and the last, q^2, is the probability that it will not happen at all. For three trials the expression gives

$$p^3 + 3p^2q + 3pq^2 + q^3$$

The successive terms give the probabilities that the event will occur three times, twice, once, or not at all.

In consideration of dispersal, p may represent any basic probability such as (a) the probability of successful dispersal of a single individual, or (b) the probability of dispersal from a given population in any unit of time such as a year. In case (a), t may be taken as the number of individuals in a population. In case (b), which is perhaps the more enlightening model for considering biogeographic events, t should be taken to represent the number of years (or other time units) involved.

In either case, in biogeographic problems the pertinent values of t will commonly run into millions. Expanding the expression

$$(p+q)^{1,000,000}$$

is a formidable undertaking, but fortunately this is not necessary. In these problems what we usually want to know is the probability that the event will not occur at all or that it will occur at least once. The probability of nonoccurence is the last term in the expanded binomial, and this is always simply q^t. Raising such a fraction as, say, 0.99999 to the millionth power may still seem unduly formidable, but this can easily be done with a table of logarithms to a sufficient number of places. The probability of occurence one or more times is necessarily one minus the probability of nonoccurrence, or $(1-q^t)$. In some cases, probability of occurrence just once or of occurrence more than once may also be of interest, and the binomial gives fairly simple formulas for these values, too. In all, the following symbols and formulas are useful:

p = probability of dispersal analogous to a single trial, e.g. for one individual or for a population in one year, one fruiting season, etc.

$q = 1 - p$ = probability of nonoccurrence, of failure in dispersal

t = a quantity analogous to number of trials: the number of individuals in the population, number of years or seasons, etc., as appropriate

a = total probability of nonoccurrence = q^t

b = total probability of occurrence at least once = $1 - q^t = 1 - a$

c = total probability of occurrence just once = $tpq^{(t-1)}$

d = total probability of occurrence more than once = $(1 - q^t) - tpq^{(t-1)} = b - c$

For accuracy, the values of a, b, c, and d must of course be worked out by formula from separate values assigned to t and to p (the value of q being fixed by that of p). Since, however, the total probabilities sought tend to depend on relative values of t and p, a rough and ready approximation can be worked out more simply from the single value representing the product of these numbers, tp. Such approximations have so much bearing on judgment of biogeographic probabilities that a brief table is submitted (Table 6–7).

TABLE 6–7

Corresponding values of a, b, c, and d for fixed values of tp.
(See text for meanings of symbols.)

tp	0.001	0.01	0.1	0.5	0.7	1.0	1.5	2.0	5.0	10.0
a	0.999	0.99	0.90	0.61	0.50	0.37	0.22	0.135	0.007	0.00005
b	0.001	0.01	0.10	0.39	0.50	0.63	0.78	0.865	0.993	0.99995
c	0.001	0.01	0.09	0.30	0.35	0.37	0.33	0.270	0.034	0.00045
d	0.000	0.00	0.01	0.09	0.15	0.26	0.45	0.595	0.960	0.99950

These values are close enough as basis for general judgment for values of p from 0.1 to about 0.00000001, although there may be an error in the last place for extreme values of p. For values of tp smaller than 0.001 or larger than 10, the values of a, b, c, and d so closely approach either 1 or 0 that in most cases they can be taken as virtually equal to one or another of those extremes, although it is well to remember that these probabilities never literally and precisely amount either to certainty or to impossibility. The interesting value $tp = 0.7$ is included, because this is approximately the point at which nonoccurrence and occurrence (at least once) become equally probable.

Two simple examples illustrate use of the table:

(1) If the chance that a species of plant will be dispersed across a barrier in any one fruiting season is one in a million, what is the chance over a period of a million fruiting seasons?

$$p = 0.000001 \text{ (given)}$$
$$t = 1,000,000 \text{ (given)}$$
$$tp = 1$$
$$b = 0.63 \text{ (table)}$$

The chance is approximately two in three, or odds are about two to one—a noteworthy probability for what would seem to the observer of one or a few seasons as a virtually impossible event.

(2) If the chance of crossing a particular barrier for a given group of animals is one in a million in any one year, how long a time must elapse before the chances of crossing or not crossing are equal?

$$p = 0.000001 \text{ (given)}$$
$$tp = 0.7 \text{ (as shown in table for } a = b = 0.50)$$
$$0.00000t = 0.7$$
$$t = 700,000 \text{ years—geologically a brief time}$$

In a given case, the subjective statement that an event is "improbable" may correspond (if the statement as made is true) with $p = 0.49$, or with

$$p = .0000000000000001$$

The implications are extremely different for these two values, which nevertheless can hardly be distinguished in many arguments about dispersal. Even though the values cannot (at present) be calculated exactly for particular cases, it is possible to judge their orders of magnitude under given circumstances.

Source

(1952) Probabilities of Dispersal in Geological Time, *Bulletin of the American Museum of Natural History*, Vol. 99, pp. 163–176. (The excerpt here is from pp. 171–173.)

Species Density of Recent North American Mammals

Comment

The total number of species (or sometimes higher taxa) of a given group of animals or plants in various geographic regions is a biogeographic factor quite different from any previously exemplified in this book. This aspect of biogeography was already receiving some attention a century or more ago, but as in so many other biological subjects it has become increasingly quantified in the last twenty years or so. Even for this one subject within the general field of biogeography there are now a number of approaches different in scale, aims, quantitative methodology, and presentation. Among these I think density contouring to be one of the most interesting, elegant, and readily understood in graphic form.

Perhaps I have overlooked earlier examples, but the earliest published use of species density contour maps known to me was in a short note published by Stehli and Helsley in 1963 (citation below). They published such maps for Recent planktonic foraminifera and for certain Permian fossil shells (orthotetaceid brachiopods). Their aim was to locate the Permian north pole on the hypothesis that the Permian brachiopods would have the lowest number of species near that pole and the highest number near the equator. The rationale was based on the long well-known fact that many groups of Recent organisms do have that sort of distribution of species densities. Deduction from their data and hypothesis put the Permian poles and equator in approximately their present position, but, as they pointed out, the paleomagnetic data then available put the Permian latitudes about 90° from the present ones. The conclusion was "that it may be necessary to consider a model which does not require coincidence of the rotational and magnetic poles." That interesting suggestion has been either ignored or rejected by most later students of biogeography and of paleomagnetism.

The study of species density here reprinted in part was already well advanced when the note by Stehli and Helsley was published, and this study was published a few months after theirs. Its use of the contour-map method was independently hit upon, had different aims, and was more complex. Its results were also differently and more extensively analyzed. The description of the method and most of the results and their analysis are here reprinted without essential change. A few pages considered as of less present interest or value are omitted. They include some discussion of high species density in the tropics, one of the population phenomena earliest noticed and most extensively discussed. That discussion has not abated since 1964, and in my opinion it has not even now produced a clear and adequate causal explanation of this phenomenon. I therefore here simply refer to a more recent review of the subject by May (1975).

Text

The basic data for this study consist of records of species of Recent mammals present in quadrates of equal area covering continental North America. The method was to take a map on Lambert's azimuthal equal-area projection and to superimpose a rectangular grid making quadrates 150 miles on each side (22,500 square miles each) not oriented with respect to physiographic features or other known zoogeographic factors. A list of species of recent mammals was compiled and the presence or absence of each species noted for each quadrate. The species lists and distributions were based on Hall and Kelson (1959), with modifications mentioned below. Only continuous mainland areas and species occurring on them were included. Insular distributions have their own special interest, but the fact that they are special would make their combination with mainland data confusing. The region covered is all of North America, including Central America to the Panama–Colombia border. The total number of quadrates is 453.

Quadrates of this size are suitable for some kinds of zoogeographical analysis and not for others. For almost any purpose using quadrates it is desirable that most or all species should occur in more than one and preferably in many quadrates, and that sets a rough upper limit of useful area. Although there are a few apparent exceptions for species the validity or extent of which is not in fact established, species of noninsular mammals in North America, at least, have areal distributions definitely and for the most part greatly in excess of 22,500 square miles. It is further necessary that quadrates be small enough for adequate resolution of the phenomena being investigated. Quadrates 500 miles on a side, for instance, probably would not resolve important zoogeographical changes adequately, even on a continental scale. For certain organisms and certain types of ecological studies, resolution may require quadrates as small as a square meter, or even smaller. Localized zoogeographic studies of vertebrates might utilize quadrates on the order of one to a hundred square miles, but that is too small to be practicable (at present) or desirable for studies of continental magnitude. For such small quadrates, actual presence or absence of a species would in many instances be variable, ephemeral, accidental, due to highly local factors, or indeterminate. A quadrate would not really represent the whole and long-term fauna of an area. Moreover, and equally important, our distributional data for species as a whole over the continent are certainly not accurate within 10 miles.

Although the choice of 150-mile quadrates was somewhat arbitrary, the order of magnitude was determined by the preceding considerations. For species density, at least, the subject of the present paper, the results seem to endorse the choice empirically. Quadrates much smaller that 150 miles on a side would have obscured parts of the overall pattern by local, more or less random fluctuations and imperfections of the data. Quadrates much larger would have lacked significant resolution. After the fact, the results do suggest that if data sufficiently precise had been available, quadrates somewhat smaller might have given desirable increase of resolution without undue loss of pattern. It is, however, dubious whether that refinement would repay much increase in the labor of tabulation.

The data, as modified from the compilation by Hall and Kelson, have a number of known and not wholly corrigible deficiencies. Application of more meaningful criteria for biological species has markedly reduced the number of nominal species of North American mammals in recent years, but the number listed by Hall and Kelson is still certainly too large, as those authors recognize. I have gone beyond them, for instance and most obviously in eliminating almost all the ridiculously defined "species" and "genera" of bears that they still list. (Only three species and one genus, as *reasonable* defined, are present in North America.) In a number of other instances, also, I have united species that Hall and Kelson list as separate, but I have done so only when a student of the group has stated that the synonymy is probable, at least. There are numerous other cases, especially in groups not recently revised, in which the distributions and stated diagnoses make synonymy extremely likely on the face of things. I have not, however, eliminated such probable synonyms on my own authority. My list thus is also surely still too long, although somewhat shorter than that of Hall and Kelson. There is, nevertheless, no reason to believe that unrecognized synonymy is more common in one *region* than in another. The sort of general patterns and conclusions here discussed are therefore probably not significantly falsified by this factor. Man and introduced mammals have not been included in my data. As noted above, entirely insular species are also omitted. The total number of species on my list is 670.

Most of the distributional patterns have now been worked over in considerable detail by several generations of mammalogists, and Hall and Kelson also checked these in numerous and large collections. There are, of course, exceptions, for instance, in some Central American species known only from one or a few scattered localities, but in general the distributions are probably now quite accurate at the scale of resolution implicit in this study and within methodological limitations. I have modified Hall and Kelson's data on areal distributions only by elimination of a few extensions obviously recent and under human influence, and of course by uniting distributions of species which they separate and I unite.

There are limitations in all the usual distribution maps for species, and these are not always understood. Such maps, like Hall and Kelson's, usually draw a smoothed limiting line around peripheral collecting records and indicate continuous distribution within that limit. Actually, of course, the limits are highly variable in time, and even if perfectly accurate for some one time (an impossibility in itself) they would not be quite the same at any other time. Moreover, the limits indicated for different species are not necessarily, or even usually, contemporaneous. An extreme but otherwise characteristic example is that of the bison, whose distribution has varied enormously and continually ever since they reached North America in the Pleistocene. The present, 1964, distribution obviously has little zoogeographical significance. The nominal distribution here used for the bison is that of latest prehistoric or earliest historic times, if indeed it is even roughly accurate for any *one* time. That is certainly not contemporaneous with other distributions, most of which are for twentieth-century dates.

The implication of continuous distribution within figured limits is also often or usually false. At most it could presumably mean that the whole area was covered by ranges of individuals of the species, which must rarely if ever be literally true. In fact, individuals of a stated species may be entirely absent over the greater part of its mapped distribution, for example, animals in a discontinuously distributed environmental niche or in a single faunal zone among several in the region. It is well known to field workers that at any one locality they can never hope to observe or collect all the species whose mapped distributional ranges cover that point.

These shortcomings must, of course, be borne in mind in any interpretation of the data. In some particulars they do seem to deprive analytical details of reliable significance.

They do not, however, seem seriously to affect overall patterns and trends for the continent. As regards species density in particular, the fact that a definite and understandable pattern does emerge is empirical evidence that the data are adequate and that their interpretation is significant in spite of some coarseness of resolution and uncertainty of detail.

The species density pattern was established by counting species in each quadrate, entering the figures on the map, and contouring by isograms. The result is shown in Fig. 6–8. The contour interval (change in number of species from one isogram to the

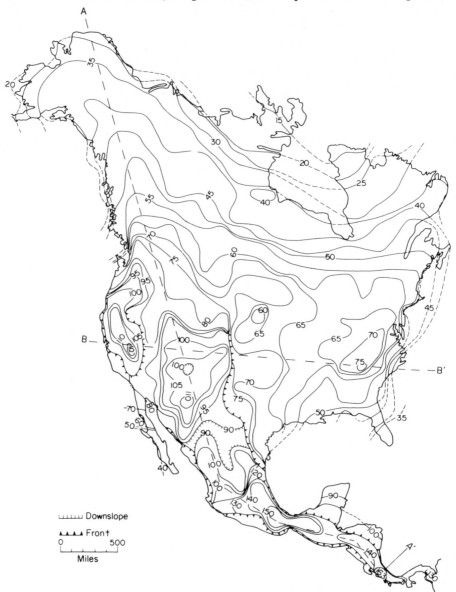

Fig. 6–8. Species density contours for recent North American mammals. See text for explanation.

next) is five for Canada and the United States. Near the Mexican border and southward it was found that species number change so rapidly that a contour interval of less than ten species was impractical on this scale. That difference is of some interest in itself. Since the absolute numbers of species are also higher in Mexico and Central America, the rate of change in terms of percentage of species present tends to be more nearly the same, although some difference persists. In other words, the density slope (change in number of species per unit of distance) tends to be a function (but not a simple linear function) of the absolute density (number of species in any one quadrate).

That is also visible in Fig. 6–9. In Alaska and Canada the average species density is about 50. The slopes, all positive in a southerly direction and accelerating in that direction, average about one species per 40 miles. In the United States the average density is somewhere between 90 and 95 species, and the slopes (in both directions) average about one species per 25 miles. In Mexico and Central America the average density is about 130, and the slopes, in both directions but predominantly positive toward the south, average about one species per 20 miles. (These approximate figures apply only to the section given in Fig. 6–9, along line A–A' of Fig. 6–8.) It is, however, evident that the relationship is loose and irregular.

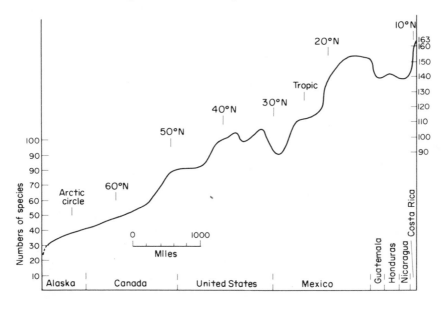

FIG. 6–9. Species densities along line A–A¹ of Fig. 6–8.

The most obvious overall feature of the map is a marked increase in species density from north to south, or equatorward. Another, more localized set of features seems to be correlated with topography, as it involves relative highs in mountain and plateau regions and relative lows on plains and in basins. A third possible but dubious regularity is what seems, as a first approximation, to be a tendency for species densities to increase from the coast inland. Also dubious but possible is some trend for increase from east to west. A striking special feature, related primarily to topography, is the occurrence of some lines of extremely abrupt change in density—the lines I have called "fronts" and indicated by a special symbol in Fig. 6–8. Finally, there is an evident tendency for large peninsulas to have fewer species than otherwise similar inland regions.

In addition to these apparent widespread tendencies, there are numerous minor curves and quirks in the contours, of the sort considered anomalies or "noise" in another recent study involving contouring of density of taxa (Stehli and Helsley, 1963). It would, of course, be unsound to qualify variations as "noise" simply because they fail to agree with a hypothesis, such as that of a polar–equatorial gradient in density of taxa. Some of the unresolved irregularities are doubtless due to deficiencies and inaccuracies of data, in which case they could reasonably but allegorically be considered noise in the sense of not containing correct information about an objective natural relationship or condition. I believe, however, that in our case (and perhaps also in that of Stehli and Helsley) most or all of these apparent irregularities do in fact contain information that is below the clear resolving power of the grid or that we are simply too ignorant to interpret. In other words, they are probably systematically caused by local conditions unknown to us or not evident at this scale.

The North–South Gradient. Mammalian species densities vary enormously in continental North America, from 13 per quadrate in one of its northernmost areas (Melville Peninsula, latitude 70°) to 163 in one of the southernmost (Costa Rica, latitude 10°). Between these extremes there is a clear but interrupted and irregular gradient. In the colder parts of the continent, from the northernmost continental land approximately to the Canadian–United States border, this gradient is fairly regular. South of that Arctic–Cold Temperate zonal gradient, more or less between latitudes 50° and 30°, that trend is absent. In 18 quadrates spanning the continent approximately along the 45th parallel, the average species density is 72. In 15 quadrates similarly arrayed along the 30th parallel, it is 68. That might, indeed, suggest a feeble trend *opposite* to that farther north, but from further examination it appears that there simply is no general north–south trend in this area. That is certain as regards any trend significant in the data as here tabulated. This area, roughly the United States, has a definite but quite irregular pattern dominated by other trends and local factors and devoid of the latitudinal gradient generally expected for such data.

To the south of that area, mainly in Mexico and roughly between latitudes 30° and 20° (i.e. in the subtropical–high tropical zone), the pattern is also quite irregular and reflects other factors as well, but there is again a definite and strong north–south gradient of increase in density. Still farther south, through Central America, the continental land is so narrow, is so abruptly varied in physiography, climate, and vegetation, and changes so little in latitude that any overall trend that might exist is concealed by local perturbations. All one can say is that a north–south trend cannot be seen in our data for this region.

Fig. 6–9 is a profile along line A–A′ on the map given in Fig. 6–8. It runs from extreme northwestern Alaska, an area of very low although not minimal species density, to the quadrate of maximal density, in Costa Rica. The striking overall increase equatorward is evident, and so are the Arctic–Cold Temperate zonal gradient through Canada, the much steeper Subtropical–Tropical zonal gradient in Central Mexico, and the absence of a clear trend in Central America. The irregularities in Central America and in the United States (Midtemperate subzone) correlate with topography and confuse or conceal any possible underlying latitudinal trend. The section in Fig. 6–12 follows the 100th meridian, along which there are few marked topographic irregularities north of Mexico.

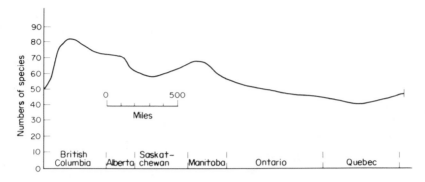

Fɪɢ. 6–10. Species densities along the 50th parallel, based on contours in Fig. 6–8.

The trend for increase from the Arctic down to the 50th parallel, or somewhat beyond, is again evident, but here it is clear that there is no such trend on southward to beyond the 30th parallel. It is impossible to make a similar check in Central America, where there are not extended possible section lines without great topographic irregularity.

Some details as to just what is going on as faunas change from north to south and east to west are deferred for later study. Here is given only brief mention of some points that could bear on attempts to explain the latitudinal gradient. Gradients for higher taxa tend to be similar to those for species, but the generic, familial, ordinal gradients are generally flatter, both in absolute numbers and in terms of percentage change. Individual orders and families frequently do not follow the overall trends. For example, the orders Insectivora and Artiodactyla and the rodent families Geomyidae and Heteromyidae reach maximum diversities at various latitudes down to the Tropic of Cancer and become *less* diverse more to the south. The groups of which this is true are without exception old in the Holarctic region and newcomers in the Neotropical region, if they reach it at all. Some old Holarctic groups (e.g. Vespertilionidae, Sciuridae, Cricetidae), however, do maintain or even increase their diversity in the present tropics. Almost all of these basically northern groups have a definite increase in diversity from north to south in Canada, even though the trend may not be maintained through the United States and may be reversed in Central America.

The increase in tropical diversity as compared with the Temperate zone does involve some old northern groups, but it is primarily due to old southern groups, most of which even now do not extend far from the tropics: opossums, phyllostomatid bats, monkeys, edentates, and caviomorph rodents. Thus the numerical expansion of taxa of mammals in the tropics is not simply an inflation of numbers of species within more or less the same higher categories but marks definite changes in overall taxonomic composition and ecological characteristics of the faunas.

What actually exists as regards species densities in these mammals is not a single polar–equatorial gradient. There is first a northern gradient that is largely (but not altogether) a simple expansion of members of species within faunas of similar basic familial and ordinal composition. This reaches an apparent equilibrium in mid- to warm-temperate regions, where there is no further north–south trend in diversity. In the temperate–tropical transition there is again—and here much more steeply—an increase in species densities. Here this reflects an extensive (but not total) change in faunal type as regards both higher taxonomic categories and ecological makeup. Thus, what is to be explained is not a gradient but two gradients of different kinds and the absence of gradient between them.

Species Density Fronts. In western United States and through Central America there are lines along which changes in species numbers are so abrupt and great that they cannot practically be represented by separate contours with the intervals and at the scale here used. The change in species numbers between adjacent quadrates is here several times the selected contour interval (5 in the United States, 10 in Central America). These lines, called "fronts" and given a special symbol in Fig. 6–8 for convenience, are not qualitatively different from other density gradients but simply represent extremely steep gradients. They are in every instance topographically correlated, coinciding (within limits of resolution of the data) with abrupt changes in relief. At no point does a north–south or climatic gradient reach this degree of steepness. The central Mexican temperate–tropical gradient most nearly approaches it.

East–West Gradients. East–west sections anywhere north of central Mexico all show a pronounced lateral asymmetry, with average species densities decidedly higher to the west than to the east, as exemplified in Figs. 6–10 and 6–11. (From central Mexico southward, no such tendency can be seen, but there are no east–west lines long enough to involve it there.) The asymmetry is influenced by topographic factors, the west having higher relief in general. In the Canadian section (Fig. 6–10), the species high in Alberta and British Colombia corresponds with the Rocky Mountains. The species high in Manitoba is also in an area of relief somewhat greater than immediately to the west or east, but the rise in species density here does seem more than would be expected from topography alone. Even the slighter rise in eastern Quebec approximates a local increase in topographic relief. Nevertheless, the rather steady rise from central Quebec westward through Ontario is not clearly related to topography.

In Fig. 6–11 the rises and falls in species density correspond closely with increases

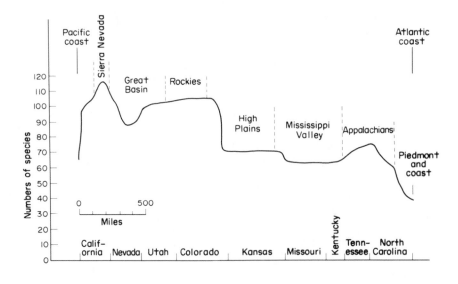

FIG. 6–11. Species densities along line B–B¹ of Fig. 6–8.

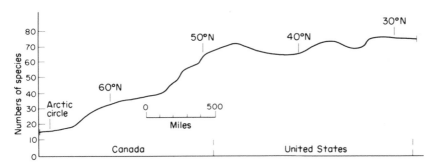

F<small>IG</small>. 6–12. Species densities from the Arctic to the Mexican border along the 100th meridian, based on contours of Fig. 6–8.

and decreases in topographic relief, as previously noted. Nevertheless, the species densities seem to be greater in the center and west than in the east for equivalent amounts of relief. The western high plains have species densities as high as those of the Appalachians even though the relief is much less in the former.

It is thus quite possible but not fully established that there is an east–west gradient in species density superimposed on and not clearly separable from the topographically correlated gradients. If the east–west gradients exists, it would be correlated roughly and *negatively* with mean annual precipitation. That is probably contrary to subjective expectations of most ecologists, who have sometimes assumed that drier climates are more rigorous and that more rigorous environments will have fewer species. There is a possible explanation in terms of topography plus precipitation. Relief in a more arid region does in general involve greater variety in climates, microclimates, and hence more generally in ecological niches. It is quite apparent that environments are more varied in, say, a south-western desert range with 3,000–4,000 feet of relief than in an Appalachian area of equal relief. It is, however, hard to see how any such factor could account for the relatively high species densities (average about 71 per quadrate) in high plains areas of *low* relief in the United States west of the 20-inch isohyet and east of the mountain front. A possibility is that increased speciation in mountainous areas has affected adjacent plains by subsequent expansion of originally montane species. It is of course known that montane species do sometimes spread to plains, and vice versa, but whether this does generally increase density of species has not been investigated.

However, that may be, the species density pattern gives no evidence of a positive correlation with precipitation in equivalent latitudes but is definitely opposed to the existence of such a relationship.

Source

(1964) Species density of North American Recent Mammals, *Systematic Zoology*, Vol. 13, pp. 57–73. (The parts here were on pages 57–63 and 70–72.)

Other References

MAY, R.M. (1975) Patterns of Species Abundance and Diversity, in, *Ecology and Evolution of Communities*, M.L. CODY and J.L. DIAMOND, editors, Belknap Press, Cambridge: pp. 81–120.
STEHLI, F.G., and C.E. HELSLEY (1963) Paleontologic Technique for Defining Ancient Pole Positions, *Science*, Vol. 142, pp. 1057–1059.

Historical Analysis of a Continental Mammalian Fauna

Comment

The history of the South American land mammals through the Age of Mammals has been studied and discussed for some three generations. A treatment of its present status is given in a book in press when these lines were written and it is cited below. Rather than summarize that book or reprint one of the shorter and earlier treatments that the book expands and replaces, I here include a historical analysis of mammalian faunas on the other island continent, Australia. The fossil evidence for Australia is much less extensive than for South America, and inference for the Australian faunal history depends largely, although not entirely, on the Recent fauna and that of the Pleistocene, the epoch just preceding the Recent.

Although studies of Australian mammals have become increasingly numerous since 1961, few of them directly affect the data or interpretations of this paper. Present classifications now include one living Australian family (Burramyidae) believed to be extinct when this paper was written and some of them put at least one living genus (*Tarsipes*) in a family of its own (Tarsipedidae) although it was referred to a broader family (Phalangeridae) in most previous classifications. Pleistocene fossils from New Guinea somewhat increase resemblance of Papuan and Australian marsupial faunas. Some recent students divide the classical order Marsupialia into two or several distinct orders, but do not agree as to what those orders should be or include. There have also been some not very important proposed changes in the classification of bats and rats. The pre-Pleistocene evidence has been considerably expanded, but without carrying it much farther back in time or greatly modifying its general nature.

In my opinion, none of those changes in the last vicennial or so require really significant modification of the inferred history here presented. There has, however, been one important change in knowledge of the geographic background that affects the interpretation of the biogeography. It is now established as reasonably probable that early in the Age of Mammals (Cenozoic) Australia was still near eastern Antarctica, that it then (probably in the Eocene) drifted away on its own and moved into approximately its present relationships with southern Asia and the East Indian island arc in about the early Miocene. With this geographic background it is still most probable that marsupials were present in Australia in the late Cretaceous or earliest Paleocene and underwent their basic Australian radiation while isolated on that continent. They may have originated there or may have reached there from South America via Antarctica by island hopping or sweepstakes dispersal. It also remains probable that ancestral monotremes were in Australia even earlier and never occurred on other continents in the Cenozoic.

It now appears extremely improbable that the ancestors of the living Australian bats were in Australia when it started its drift northward or that they began hopping along the island arc from Asia to Australia in the early Cenozoic. They clearly did not spread from or to South America via Antarctica. The seven Australian families of Tate's classification are all ancient in Eurasia. Four of them have never occurred in the Americas as far as known. The other three (Emballonuridae, Vespertilionidae, and Molossidae) are now worldwide. Of their 14 Australian genera, none occur in North or South America and all but two are still extant in Asia and probably originated there or elsewhere in Eurasia, or possibly Africa for some of them. Moreover none of the distinctive families of bats now confined to tropical America are also present in Australia.

Thus it is now still probable almost to the point of certainty that all the ancestors of the recent Australian bats came from southern Asia, but it is now most likely that none of them reached Australia until Miocene or later times. This is also reasonable in view of the fact that further evolution *in situ* in Australia–New Guinea was mostly at the specific level and not in any case beyond the generic level.

Inferences about the history of Australian members of the rat family (Muridae) are not affected by this change in evidence for the geographic background. Recent finds and studies in Eurasia increase the probability that the Muridae arose in southern Asia in the early Miocene or little before. It is still reasonably certain that the ancestors of all the Australia–New Guinea native murids were ultimately of Asiatic origin and that they began to reach the Australian zoogeographic region not earlier than the Miocene.

The terms "native," "endemic," and "autochthonous" are sometimes used in different ways, so I here note the way in which I use them, which I believe to be most precisely distinctive and therefore best in biogeographic use. Native taxa or populations of organisms in a given place or region are those living and reproducing there and not having been introduced by direct or indirect human action. Endemic taxa are native taxa that occur only in the specified region. Autochthonous taxa are native taxa, whether endemic or not, that originated in the specified region.

In this paper I used the toponym New Guinea for the whole of the immense island now divided politically into Papua New Guinea to the east and Irian Barat (or West Irian) to the west. These usages of "New Guinea" and "Papuan" are historically correct in English and there are no other usable name and adjective for the island as a whole. (The Malay name "Irian" could be used in both ways, but it is not generally understood by English readers and it is politically objectionable to citizens of Papua New Guinea.)

The Australian Region is here assumed to include New Guinea and some adjacent small islands, although New Guinea is zoogeographically a distinct subregion within that region. A recent (1977) paper about the western boundary of that region was mentioned on p. 229, above, and is cited there. It is not reprinted in the present volume, but Fig. 6–13 here is a modified version of a map compiled for that study.

This reprint has been somewhat abbreviated by omission of parts now of only historical interest and others more technical than seems necessary here.

Text

Two of the usual characteristics of insular faunas, impoverishment and endemicity, are evident in the general makeup of the Australian fauna as summarized in Table 6–8. Both

TABLE 6–8

Australian native recent mammals

	Families		Genera	
	No.*	Endemic†	No.*	Endemic†
Monotremata	2	100%	2	100%
Marsupialia	6	100%	47	100%
Chiroptera	7	0%	21	29%
Rodentia	1	0%	13	77%
Carnivora	1	0%	1	0%
Totals	17	47%	84	67%

*The numbers refer to inhabitants of continental Australia and Tasmania (and not the whole Australian Region).

†Groups are counted as endemic if confined to the region east of the Asiatic continental shelf.

factors vary with categorical level. The numbers of species (not given in the table) and of genera are not much smaller than would be expected from the size and environmental ecology of Australia. The number of orders, only five, is, however, even below that of Madagascar, with six. On other continents it ranges from seven in Europe to twelve each in Asia and Africa.

The idea of Australia as the land of marsupials and monotremes has been so much stressed that even some technical studies have overlooked the fact that Australia does have a rich placental fauna. In fact over half the native orders and families of Australia and nearly half the native genera are placentals. The idea is, however, correct in a way, because the placentals are either of comparatively recent introduction or of only two broad adaptive types (bats and rats). The great majority of ecological niches for land mammals are, indeed, filled by marsupials. The frequent insular characteristic of faunal imbalance is hardly evident in Australia because the ancient, spectacular marsupial radiation has produced an essentially balanced fauna of land mammals within that one order. The broad outlines of placental zoogeography in the Australian Region are nevertheless more complex and in some ways more interesting than for the marsupials. Endemicity varies not only categorically but also taxonomically. It is complete from specific through subclass levels for monotremes and through superfamily level for marsupials. It is nearly but not quite complete for species of rodents, and near three-quarters for genera, but the single family is not endemic. For bats, specific and generic endemicity is considerable but is decidedly less than for rodents, and none of the seven families is endemic. The one (placental) carnivore is not endemic at any level above the species.

The differences in endemicity reflect the fact that ancestral members of the various orders reached Australia at different times and in different ways, and that rates of evolution have also differed. In these respects each order is a separate case, and they will be separately discussed, with the sequence reversed from that usual in classification.

Carnivora. It is well known that the only placental carnivore in Australia at the time of European discovery was the dingo. It interbreeds freely with domesticated dogs and resembles them more than it does any wild species. The general assumption is that dingos are feral descendants of domesticated or semidomesticated dogs introduced by aboriginal humans, although the archaeological evidence seems to be inadequate or ambiguous. If the assumption is correct, which is probable, this would be a unique case of a population that has passed through a phase of domestication and some degree of artificial selection and then had fully returned to the status of a wild species through many generations in isolation. (That is not really true of, for example, the "wild" horses of the American west.) Such a history may have had unusual genetic and adaptive consequences. Unfortunately it is probably too late to investigate that possibility adequately.

Rodentia. The numerous species of native rodents in the Australian Region all belong to the family Muridae. They have been studied in most detail by Tate (especially 1951), and with some later changes his data are the main basis for the following discussion. A few Pleistocene or early Recent fossils are known, but as far as yet studied they add virtually nothing of significance. From a historical point of view the recent murids can be divided into four broad groups:

(I) local members of the very widespread genus *Rattus*;

(II) old Papuan genera: nine genera on New Guinea, two of which also occur in Australia;

(III) the *Pseudomys* group: eight genera, almost confined to Australia;

(IV) the subfamily Hydromyinae: nine genera in New Guinea, one in Australia, and one common to both.

These four groups have had different histories, and each is rather complex within itself.

(I) *Rattus.* The now ubiquitous species *R. rattus* and *R. norvegicus*, along with *Mus musculus*, were introduced into the Australian Region after the European discovery. Rats of the *R. exulans* group are widely distributed in the Pacific in a way consistent with prehistoric (i.e. pre-European) spread by native canoes and difficult or impossible to explain otherwise. It is in accord with that mode of introduction that the group is well established in New Guinea, which is more in the prehistoric seafaring lanes, and is comparatively rare and marginal in Australia.

Besides those introductions, late and early, there are seventeen species of *Rattus*, nine in Australia, five in New Guinea, and three in both, that are endemic to the region and seem also to be autochthonous in it. Their differentiation there implies considerable antiquity and prehuman immigration of their ancestry. Tate places them in two of his "divisions" and eight of his "groups" of species of *Rattus*, but all could have differentiated from two or less likely from one ancestry migrant through the East Indies. From the degree of differentiation, arrival in New Guinea and hence in Australia could hardly have been much later than early Pleistocene. There has been some limited later Pleistocene or early Recent intermigration between New Guinea and Australia, with *R. ruber* and *gestri* probably moving from Australia to New Guinea and *R. leucopus* probably in the reverse direction.

(II) "Old Papuans." *Uromys* and *Melomys*, common to Australia and New Guinea, and *Xenuromys* and *Pogonomelomys*, confined to New Guinea, comprise a group (the "*Uromys* group") of related rats all of which could well have been derived within New Guinea from a single ancestry near or in *Rattus*. Generic differentiation suggests that the ancestry probably reached New Guinea not later than the Pliocene. *Uromys* and *Melomys*, both of which have speciated widely in New Guinea, almost certainly originated there and spread later to Australia. *Melomys* migrated earlier, for it is now widespread in northern and eastern Australia and has developed new species there. *Uromys caudi-*

maculatus, not specifically distinct from its Papuan relatives and confined to northeastern Australia, must be a comparatively recent migrant. Within this group only lowland and chiefly rain-forest forms spread from New Guinea to Australia, and there seems to have been no back migration.

There are five other old Papuan genera endemic in New Guinea: *Mallomys*, *Hyomys*, *Anisomys*, *Pogonomys*, and *Macruromys*. They are not especially related among themselves, aside from all being murids, and seem to have no special relatives elsewhere. They are even more distinct than the *Uromys* group and must be rather old, Pliocene, at least, and probably Miocene, in New Guinea and may represent several different invasions through the East Indies.

(III) "Old Australians," or Pseudomyinae. These highly varied but related members of the *Pseudomys* group are the most common and characteristic native rats of Australia proper. They include the following main adaptive types:

(1) Medium-sized to small, relatively unspecialized. Most abundant and diverse. *Pseudomys* (including *Thetomys* and *Gyomys*, sometimes given generic rank), *Leggadina*, and *Zyzomys* (including *Laomys*).

(2) Large, gregarious to colonial, nest-building, with very large ears. *Leporillus*.

(3) Medium large, with short tail and slender feet. *Mastacomys*.

(4) Small to moderate in size, with long ears, tail, and feet, saltorial. *Notomys*.

(5) Very large, arboreal, with long, hairy tail. *Mesembriomys*.

(6) Large to medium in size, eyes large and ears long, hopping, with long hind feet. *Conilurus*.

Among themselves these related forms represent a well-marked adaptive radiation which has produced a variety of animals convergent toward several different groups of rodents and rodentlike animals of other continents: true rats and mice (*Pseudomys* and allies), voles (*Mastacomys*), kangaroo rats or jerboas (*Notomys*), squirrels (*Mesembriomys*), and rabbits (*Conilurus*). In addition, one group (*Leporillus*) is fairly unique, not closely convergent toward any other rodents. Although this is a broad gamut for one small group of genera, it is far from occupying all the ecological niches of rodents on other continents.

In spite of their diversity, all these animals could have been derived from one ancestral immigrant. The degree of divergence, most or all of which seems to have occurred within Australia, demands a remote date for that ancestry, probably no later than Miocene but probably also little or no earlier, as true Muridae are not known anywhere before the Miocene. The ancestor must have been a very primitive murid, perhaps before distinct differentiation of the Murinae. Recognition of this group as a distinct subfamily, Pseudomyinae, seems warranted.

The pseudomyine ancestors were probably the first rodents ever to reach Australia proper. Their adaptive radiation then occurred there in the absence of any placental competitors. It seems possible that they replaced some earlier rodentlike marsupials, and it will be interesting to see whether such marsupials appear in the now unknown Australian Oligocene fauna.

It seems probable that the pseudomyine ancestors were, like the "old Papuans," early migrants to and through New Guinea. If so, the lineage either died out in New Guinea after reaching Australia or is just possibly represented by the Papuan endemic genus *Lorentzimys*. Tate believed that *Lorentzimys*, a small mouselike or *Leggadina*-like murid, belongs in the *Pseudomys* group, although the not very detailed resemblance could be convergent. If *Lorentzimys* really is especially related to *Pseudomys* (and *Leggadina*), it is at least probable that it is a survivor of ancestral Papuan pre-pseudomyines as that it was a migrant from Australia to New Guinea.

A single species of *Conilurus* has spread, certainly in late Pleistocene or Recent times, from northern Australia to southern New Guinea.

(IV) Hydromyinae. This peculiar subfamily is characterized by basined cheek teeth and reduction or loss of the posterior molars. Of its eleven adequately established genera, nine are confined to New Guinea. Another, *Hydromys*, has a species, *H. chrysogaster*, common to New Guinea and Australia. Although the species is now widespread in Australia, the Australian and New Guinean populations are not clearly subspecifically distinct and must represent a quite recent spread from New Guinea.

Xeromys, a divergently specialized member of the subfamily, is confined to a small area in Queensland, Australia. Tate (1951) considered it "a localized invader from New Guinea during middle Pleistocene time or earlier." That is possible but improbable. The genus has no special, close relatives in New Guinea, and differentiation from surviving New Guinea hydromyines would probably have taken a longer time than since middle or even early Pleistocene. Its distribution suggests a relict rather than a localized invader, and it is probably an early offshoot of Papuan hydromyine ancestry that reached Australia in the Miocene or Pliocene and became divergently specialized there.

Everything points to the conclusion that the Hydromyinae, as such, originated not in Asia, in the Philippines, or along the East Indian island stepping stones but in New Guinea itself. The remote ancestry, necessarily near the base of the Muridae and possibly even in premurid cricetids, doubtless was Asiatic and migrated through the East Indies, but hydromyine specialization was Papuan. In New Guinea the group had an adaptive radiation analogous to that of the Pseudomyinae in Australia but different and more limited in ecological scope. About half of the genera, including those apparently most primitive (e.g. *Leptomys*) are terrestrial, some of them rather shrewlike. Adaptation to subaquatic and aquatic habits occurred within the subfamily radiation. That adaptation added other convergences, notably toward muskrats and some other aquatic rodents.

Summary on Muridae. The murids of the Australian Region clearly did not reach there over a land bridge but by island-hopping down the East Indian chain. From some time in the Miocene (perhaps but less probably even in late Oligocene) there was probably continual, intermittent drift of murids down that chain. There must have been marked attenuation, with fewer originally Asiatic lineages represented at increasing distances from the mainland. There was local differentiation on various islands along the route, and such differentiation was passed on, in part, to the next stepping stone and eventually to Australia, where final and in some cases most extensive differentiation occurred. Most and probably all of the Australian stocks passed through New Guinea, with varying degrees of differentiation there before spreading onward to Australia. In some instances that differentiation was less than specific; in others it was subfamilial. Although continual, the migrations may be analyzed into three successive waves:

(A) Oldest wave, mainly or wholly Miocene. There was some basic radiation in New Guinea but this wave of invasions was evidently multiple, including up to seven but probably fewer different stocks already phylogenetically distinct before or while passing along the East Indies. Five of these stocks (the "old Papuans") remain endemic in New Guinea. Another radiated there into the hydromyinae, one lineage (*Xeromys*) of which reached Australia early and another (within *Hydromys*) late. Another stock, the Pseudomyinae, although probably of old Papuan origin, had its major radiation in Australia. A single late species reached New Guinea from Australia.

(B) Intermediate wave, probably Pliocene. This involved the probably single stock that evolved into the *Uromys* group in New Guinea. Spread thence to Australia involved first *Melomys* and later *Uromys*.

(C) Late wave, Pleistocene or perhaps in part late Pliocene. This involved perhaps only one but probably two partly differentiated lineages of East Indian *Rattus*. It radiated moderately in New Guinea and quickly spread to Australia where it radiated more extensively at the specific level. In the late Pleistocene to Recent there was some further, but quite limited, interchange both ways between New Guinea and Australia. (Human introductions, both prehistoric and historic, followed.)

Throughout, predominant movement has evidently been from smaller East Indian islands to the great island of New Guinea and then from New Guinea to the still greater continent of Australia. That apparently violates the zoogeographic principle that successful colonization tends to go from larger and more varied to smaller and less varied regions. The principle strictly applies, however, to movement into essentially closed communities. Here the predominant tendency was to move from what were, for these animals, more closed to more open communities. Moreover, the stocks involved, although not precisely in the form in which they arrived, were ultimately derived from the largest and most varied land mass of all. In the broad picture, this is another example of the usual sort of spread from central, major masses to marginal and terminal areas (see Darlington, 1957).

Chiroptera. Data for the bats of the Australian Region have also been taken mainly from Tate (1946), but with some emendations from later studies. Data on endemicity for genera and for Tate's species groups in each family are given in Table 6–9.

TABLE 6–9

Endemicity in Australian bats

Only groups that occur in Australia (and Tasmania) are considered. Aus. = Australia only. Aus. R = Australian Region (including New Guinea, adjacent islands, and East Indies east of the Asiatic continental shelf). Aus.-O = Australian Region into Oriental Region or beyond (including continental East Indies). Numbers in the table are of genera and of species groups as in Tate (1946), slightly modified.

Families	Genera			Species		Groups
	Aus.	Aus.R.	Aus.–O.	Aus.	Aus.R.	Aus.–O.
Pteropidae	0	3	2	0	7	3
Emballonuridae	0	0	2	1	1	1
Megadermatidae	1	0	0	1	0	0
Rhinolophidae	0	0	1	1	2	3
Hipposideridae	1	0	1	1	1	3
Vespertilionidae	0	1	8	1	3	8
Molossidae	0	0	1	1	0	1
Totals	2(10%)	4(19%)	15(71%)	6(15%)	14(36%)	19(49%)
Aus. R. total		6(29%)			20(51%)	
Aus. total		21			39	

The seven chiropteran families of Australia proper are all widespread in Eurasia and, for some of them, other continents, and all are old. No fossils have been described from the Australian region but all of these families date as far back as middle Eocene to middle Oligocene in Europe. The fossils are few, and in fact all these families undoubt-edly existed in the Eocene, and some may have originated in the Paleocene. They could, then, have reached Australia at any time since the early Tertiary, and there has been no differentiation of families and comparatively little of genera in the Australian Region.

The lower endemicity in comparison with the Muridae must be related to the greater ability of bats, flying mammals, to cross water barriers. There are no strongly marked ecological barriers, aside from the water gaps, on the route from southeast Asia through the East Indies to New Guinea. All the islands are in the same climatic zone and most of them have considerable resemblance or, at least, overlap in other ecological conditions. The importance of that factor for Chiroptera is evident from comparison with the Neotropical Region. Although there is now no water barrier between the Neotropical and Nearctic Regions, the ecological barrier for bats is so marked that 56 per cent of Neotropical families are endemic, as against none in Australia, and 87 per cent of the Neotropical genera are endemic, as against only 29 per cent in Australia. Bat distribution resembles that of birds and of many plants more than that of terrestrial mammals.

The Australian bats are much more diverse than the rodents, and, in sharp contrast to the latter, they represent an only moderately attenuated sampling of the whole of the original mainland Asiatic fauna. They do not fall into definite groups or waves as regards antiquity in the Australian Region but seem to have filtered in continuously throughout most of the Cenozoic. There is a wide spectrum of endemicity, which may be assumed to be approximately but only approximately correlated with antiquity in the region. Some Australian species, e.g. in *Myotis*, *Pipistrellus*, and *Miniopterus*, are very close to Indian, European, and even African species, although supposedly not quite identical. On the other hand there are a few clear-cut endemic Australian genera, notably the peculiar *Macroderma*, which is a predator on other bats. As is evident from Table 6–8, the fruit bats or Megachiroptera (Pteropidae), although less diversified, are much more divergent in the Australian Region, much higher in endemicity, than the Microchiroptera (the six other families). That suggests that the fruit bats are on an average older than the other bats in this region, although the first entry of microchiropterans may have been quite as early as this or earlier than that of the megachiropterans.

In spite of the absence of endemicity above the generic level, first colonizations could have been and probably were quite early in the Cenozoic. [See previous note.] The antiquity of the families in Europe has been mentioned, and in spite of the paucity of fossil record even a living genus (*Rhinolophus*) now present in the Australian Region is reported from the Eocene of Europe. The total number of colonizations from the East Indies into the Australian Region has not been estimated, but it must have been large, on the order of twenty or thirty, possibly even more. The basic radiations, those now at family and in good part also generic level, occurred in Eurasia. There has been little true radiation in the Australian Region, but mostly local geographic differentiation below the generic level. The center of density in the region is not on the Australian continent but on the islands from New Guinea to the Solomons, where there are more groups of bats than in Australia proper. Only two of the twenty-one Australian genera and six of the thirty-nine species groups are confined to Australia. What regional differentiation there is clearly took place mostly in the islands (in part also the East Indian islands) before the groups reached continental Australia. Nevertheless two sharply distinct groups (*Macroderma* and *Rhinonicteris*) have originated in Australia, and most groups have become subspecifically to specifically distinct since reaching there.

TABLE 6–10

Taxonomic gradients of Chiroptera between
the East Indies and Australia

(I)   Percentages of species groups of supposed Asiatic and Australian (Region) origin, after Tate (1946).

"Origin"	Celebes	Moluccas	Australia	New Guinea and Solomons
Asia	90%	64%	35%	32%
Australian Region	10%	34%	65%	68%

(II)   Indices of faunal resemblance, $100C/N_1$ for species groups between the following areas:
A, East Indies, from Asia to edge of Asiatic continental shelf.
B, islands between A and C.
C, Australia and islands on its continental shelf.

Raw data from Tate (1946).

	A–B	A–C	C–A	C–B
$100C/N_1$	58	31	31	53

For the bats, as for birds and such other animals as have spread along the whole chain, there is a gradient between the Asiatic and the Australian mainlands. Tate (1946) divided his species groups of bats according to assumed Asiatic or Australian regional origin, in a somewhat arbitrary way but using criteria of greatest distinction or differentiation, abundance, and presence of relatives toward one end or the other of the sequence. Some of his data on that basis are given in the upper part of Table 6–10. He envisioned the gradients as representing two sequences of counter-migrations, one from Asia toward the Australian Region and one from the Australian Region toward Asia, both currents becoming more and more attenuated with greater distance from their assumed regions of origin. That approach seems to be rather generally accepted, but it is certainly oversimplified. It would not be a serious oversimplification for groups that did in fact have their basic differentiation and the origins of most of their taxa either in Asia or in Australia. That is true, for example, of marsupials, with virtually all taxa of Australian origin, as compared with placentals exclusive of bats and Muridae, with virtually all taxa of Asiatic origin. As suggested in Fig. 6–14, those groups do show just the kind of overlap and opposite attenuations involved in the counter-migration hypothesis.

That hypothesis is, however, an unjustified oversimplification when applied to the bats. For them there are not two clear-cut and quite distinct centers of origin from which counter-migrations occurred. All the families, most of the genera (about 90 per cent), and an undetermined but probably large number of the species groups of Australian bats originated outside that region, i.e. in the Oriental Region or beyond. The predominant pertinent movement was not in two counter-migrations, but in one direction, from Asia toward Australia. It thus seems probable that a significant element in the Australia–Asia gradient was increasing divergence or differentiation of animals really spreading in the opposite direction (Asia–Australia). During that movement they evolved progressively, in greatest part below the generic level, away from their Asiatic ancestry.

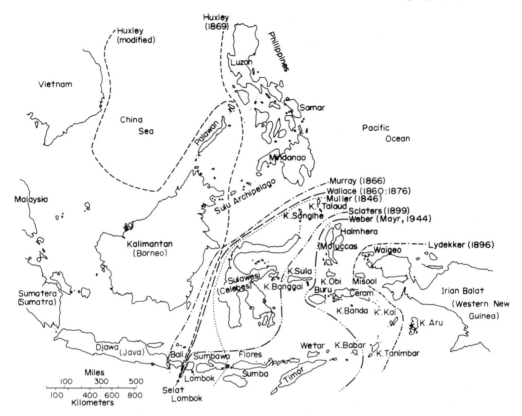

FIG. 6–13. Map of the East Indian region between the Asiatic mainland and New Guinea. The various lines shown are those that have been proposed as limits between an Oriental zoogeographic region and an Australian region. The line marked "Huxley (Modified)" follows the Asiatic continental shelf and is taken by me (Simpson, 1977) as the limit of the Oriental region. The line marked "Lydekker (1896)" is taken as the limit of the Australian region. The islands between those lines are not assigned by me to either region.

The most impressive argument against that conclusion is that faunas of intermediate islands are impoverished in comparison with either end of the sequence, and that ancestral forms or close relatives of many Australian species groups do not now occur in the Oriental Region. However, ancestors of those groups certainly did migrate through the islands even where relatives are now absent, and survival in the Oriental Region of recognizably close collaterals from the ancestry would not be expected in all or in many instances for migrations that started in the early Tertiary.

On that hypothesis, the double sets of taxonomic clines of Tate and others are in fact a single set of genetic clines, with similarities decreasing and differences increasing in the one direction: Asia–Australia. By itself, that hypothesis is certainly also oversimple. There doubtless was also some counter-migration, involving low taxa, in the direction Australia–Asia. It must also be remembered that the principal eastern center of origin and dispersal was not Australia itself but New Guinea. There must moreover have been further low-level complications by origins and dispersals in various directions around other points scattered along the sequence.

The distributional facts can be more objectively shown by measures of actually existing faunal resemblances, which do not involve any hypotheses as to the places of origin of the various taxa. That approach is exemplified in the lower part of Table 6–9. The figures show that if one starts at either more or less terminal area (continental East Indian islands; Australia and its continental islands), faunal resemblances become less with increasing distance. The measurement of that change has special interest for the particular case, but certainly the result is not surprising. The trend shown would almost inevitably appear (with different intensities) in comparisons of *any* two faunas with each other and with intermediate faunas. That result would arise in each of three models that are extreme or limiting cases and are not to be expected in pure, uncomplicated form in nature:

(1) counter-migrations between two areas originally with faunas completely different at the taxonomic levels under consideration;

(2) migration of the whole fauna from one terminal area to the other with progressive evolution along the way;

(3) local differentiation throughout a fauna that was originally uniform in and between the two areas.

In most cases of real faunas, all three factors probably act and interact, plus additional local perturbations and complications. That is doubtless true of the faunal relationships between the Oriental and Australian Regions. As regards the nonvolent land mammals, the first factor predominates in the comparatively small area of overlap (Aru to Celebes). As regards the bats, such dominance of any one factor is less clear, but the second factor is probably at least as important as either of the other two.

Marsupialia. On the scale of this study, the distribution of the Australian marsupials is much simpler than that of the Muridae or Chiroptera. It has also been the subject of more reviews as well as detailed studies. It will therefore be more briefly treated here.

In contrast with the Muridae and Chiroptera, there are some pertinent but still grossly inadequate fossil data for the marsupials. The Pleistocene to sub-Recent fauna of Australia, as on other continents, was richer than the present fauna, and it was already rather well-known in the 19th century. Information on earlier mammals is still extremely scanty. Only about a dozen forms, scattered from late Eocene or Oligocene through the Pliocene, have as yet been described and not all of those are well identified. All belong to Pleistocene or Recent families. As far as they go, they suggest that there has been some progression at generic and specific levels but that the general character of the Australian marsupial fauna did not change significantly from the mid-Tertiary, at latest, into the Pleistocene.

In conservative classification there are six living and two recently extinct families of Australian marsupials. A single species of *Phalanger* reached Celebes and thus extends into an area not now usually included in the Australian Region. With that trivial exception, all the marsupials of the Australian Region are endemic through the superfamily level. [This is not strictly in accordance with redefinition of the Australian Region, but no marsupials reach the redefined Oriental Region. Note added in 1979.] The major radiation clearly took place on the Australian continent and is surely very old. Scanty as it is, the fossil evidence strongly suggests that the families, and probably some or all subfamilies, were differentiated by the Miocene at latest, and probably much earlier. (The majority of Recent mammalian families in the rest of the world were differentiated in the Eocene and Oligocene; marsupial families may prove to be even older.)

From Australia marsupials have spread with a completely regular pattern of pro-

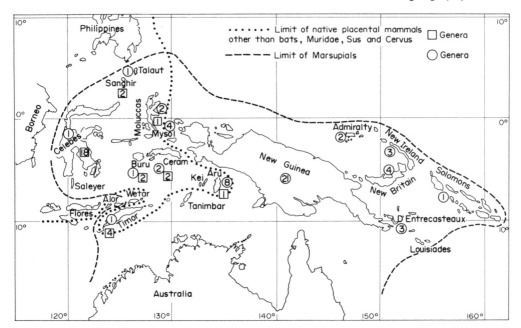

FIG. 6–14. Map of northern Australia, New Guinea, and some other islands showing present limits of native placental mammals other than bats and rats (Muridae), which reached the Australian region by island hopping, and *Sus* (pigs) and *Cervus* (deer), introduced by humans, and present limits of marsupials which spread to outer islands from Australia and New Guinea by waif dispersal. The figures in circles and squares are for numbers of native genera of marsupials (circles) and placentals (squares) on the adjacent islands.

gressive attentuation (see Fig. 6–14) as far as the Solomons to the northeast, Talaut to the north, and Celebes to the west. Four families and twenty-one genera occur in New Guinea, as listed in Table 6–11. Most of the Papuan species and six of the genera are not present in Australia proper. Spread of their ancestors to New Guinea must have begun far back in the Tertiary, and there has evidently been some intermittent interchange ever since. Even though it has developed autochthonous genera, New Guinea has apparently not been a major center for marsupial evolution, as it was for both rats and bats. Nevertheless four genera shared with Australia perhaps evolved in New Guinea and may represent back-migration in Australia: the bandicoot *Echymipera*, the phalangers ("possums" to Australians) *Phalanger* and *Dactylopsila*, and the tree kangaroo *Dendrolagus*.

Ten of the Papuan genera have also spread to other islands. They thin out rapidly and regularly with increasing distance from New Guinea. The most successful are *Phalanger*, which occurs wherever there are any marsupials in the Australian Region (and, as noted, as far as Celebes), gliding phalangers of the genus *Petaurus*, and bandicoots of the genus *Echymipera*. Significantly, all are typically rain forest groups. Ceram has an endemic genus of bandicoots, *Rhynchomeles*, which is most closely related to the endemic Papuan *Peroryctes*. It is clear that spread of all these groups to other islands has been from New Guinea and not directly from Australia. Even the Timor phalanger, *Phalanger orientalis*, although closer to Australia than to New Guinea, came from New Guinea, probably by way of Ceram, Buru, and Wetar. That species has spread from

TABLE 6-11

Marsupials of New Guinea

Dasyuridae
IA   *Phascogale* (Including *Murexia, Antechinus, Neophascogale, Phascolorsorex*)
IA   *Sminthopsis*
A   *Planigale*
I   *Myoictis*
A   *Dasyurus* (including *Satanellus*)
Peramelidae
            Peroryctes   } near Australian *Perameles*
            Microperoryctes }
IA*   *Echymipera*
A   *Thylacis* (= *Isoodon*)
Phalangeridae
XA*   *Phalanger* (including *Spilocuscus*)
A   *Eudromicia*—near Australian *Cercaertus*
IA*   *Dactylopsila*
            Dactylonax—near Australian and Papuan *Dactylopsila*
IA   *Petaurus*
A   *Acrobates*
            Distoechurus
A   *Pseudocheirus*
Macropodidae
IA   *Thylogale*
IA   *Protemnodon* (= *Wallabia*)
A*   *Dendrolagus*
I   *Dorcopsis* (including *Dorcopsulis*)

I—Extends also to nearby islands.
A—Also on Australia and probably originated there.
A*—Also on Australia but possibly originated on New Guinea.
X—Coextensive with Marsupialia on islands.

New Guinea in late Pleistocene or Recent times all over the islands that have marsupials, from the Solomons to Celebes, and also to Cape York Peninsula on Australia.

It is extremely unlikely that Australia had any nonvolant land placentals before the arrival of murids, which, as has been indicated, was no earlier than Oligocene and probably in the Miocene. By that time the basic marsupial radiation had already occurred, and its very extent supports the conclusion that the early marsupials had no placental competitors. Marsupials, abundant in the late Cretaceous and early Tertiary of North America and the early Tertiary, at least, of Europe, there faced strong and varied placental competition and became extinct by the Miocene (although one species later reinvaded the Nearctic). In South America specialized placental insectivores, rodents, and specialized placental carnivores, all absent in the earliest Paleocene of North America, were not among the original stocks derived from North America, and in their absence the marsupials underwent a partial radiation including much convergence toward those placentals. In Australia there were presumably no earliest Tertiary placentals and the marsupials underwent a radiation that was essentially complete ecologically.

The classical explanation for the absence of early placentals in Australia was that the marsupials evolved first and spread to that continent over a land bridge that was submerged before there were any placentals to cross it. Although that theory is still sometimes presented, it has long been abandoned by closer students of the problem. Marsupials and placentals, distinctly differentiated as such, apparently evolved at about the same time, and the earliest marsupials do not seem to have been significantly more primitive than the earliest placentals. The most reasonable explanation, although it is still speculative in the absence of any direct evidence, is that no land bridge was involved but that primitive marsupials were dispersed over a series of sea barriers. The original stock must have been of small, arboreal, opossumlike (i.e. didelphoid, not phalangeroid) animals such as would be particularly apt at island-hopping. Arrival of one pair or of a single gravid female could have sufficed.

Monotremata. As is well known, the monotremes belong to two families, the Tachyglossidae with *Tachyglossus* on both Australia and New Guinea and its close ally *Zaglossus* on New Guinea only, and the Ornithorhynchidae with the single Australian genus. These are extreme endemics, representing a whole subclass now confined to the Australian Region. Universal recognition that they have some very primitive characters in comparison with other (or true) mammals has sometimes obscured the fact that in other respects they are extremely specialized, about as specialized as any mammals (or reptiles). The two families are also markedly divergent from each other and must have been separated for a very long time.

A few Australian Pleistocene fossils cast no light on the history of the group. In the rest of the world nothing possibly related to them is known in the late Mesozoic or the Cenozoic.

The most probable present hypothesis on the geographic history of the monotremes is that advanced therapsids or early post-therapsids, near the arbitrary reptile–mammal line as usually drawn, reached Australia in the late Triassic or in the Jurassic, that they there gave rise to the monotremes of stricter definition, and that the latter have persisted in and have always been confined to the Australian Region. It is completely speculative, but is an interesting speculation, that there may have been a Mesozoic monotreme radiation in Australia that was mostly, yet not entirely, replaced by a later, mainly early Tertiary, marsupial radiation.

Source

(1961) Historical Zoogeography of Australian Mammals, *Evolution*, Vol. 15, pp. 431–446.

Other References

SIMPSON, G.G. (1977): Too Many Lines; the Limits of the Oriental and Australian Zoogeographic Regions, *Proceedings of the American Philosophical Society*, Vol. 121, pp. 107–120.

SIMPSON, G.G. (1980) *Splendid Isolation, The Curious History of Mammals in South America*, Yale University Press, New Haven and London.

TATE, G.H.H. (1946) Geographical Distribution of the Bats in the Australian Archipelago, *American Museum Novitates*, No. 1323, pp. 1–21.

TATE, G.H.H. (1951) The Rodents of Australia and New Guinea, *Bulletin of the American Museum of Natural History*, Vol. 97, pp. 183–430.

Index